Electromagnetic fields ⌐om a variety of natural and ⌐ sources can have adver⌐ a variety of electrical a⌐ ic systems. These effec⌐ from noise on communication lines and erratic performance in digital circuits all the way to equipment damage and personnel hazards.

This book will help those who deal quantitatively with the coupling of electromagnetic fields to transmission lines and who need effective techniques to solve these problems. Although the list of electromagnetic sources and susceptible conductors is quite long, their essential elements can be modeled in terms of their physical behavior, with a few simple mathematical forms. To do this, Smith bases coupling equations on a transmission-line theory formulation. He also develops a methodology for solving nonuniform electromagnetic field problems without restricting his solutions to plane waves.

This material is derived from Smith's research on electromagnetic coupling problems and the application of his theoretical results to actual communication problems. The book is ideal for solving coupling problems in data processing, power, telephone, or control systems. Systems engineers and circuit designers with interests in numerical solutions will find abundant application data in the form of solved examples and spectrum profiles. The electromagnetic specialist will find enough theoretical details here for easy extension to a wide variety of problems. Academicians will discover treatments in transmission-line theory and electromagnetic theory suitable for graduate or advanced undergraduate classwork. This book is readily understandable to anyone familiar with transmission-line theory and field theory.

COUPLING OF EXTERNAL ELECTROMAGNETIC FIELDS TO TRANSMISSION LINES

COUPLING OF EXTERNAL ELECTROMAGNETIC FIELDS TO TRANSMISSION LINES

ALBERT A. SMITH, JR.

Advisory Engineer
IBM System Communications Division

A Wiley-Interscience Publication

JOHN WILEY AND SONS   New York / London / Sydney / Toronto

Library of Congress Cataloging in Publication Data

Smith, Albert A., Jr.   1935-
   Coupling of external electromagnetic fields to transmission lines.

   "A Wiley-Interscience publication."
   Bibliography: p.
   1.   Electric lines.   2.   Electromagnetic compatibility.
3.   Electromagnetic fields.   I.   Title.
TK3221.S58   621.319′2   76-49504
ISBN 0-471-01995-X

Printed in the United States of America

10 9 8 7 6 5 4 3 2 1

To
Rosemarie,
Denise, and Matthew

PREFACE

Electromagnetic fields radiated by a variety of natural and manmade sources excite currents in the wires and cables of electrical and electronic systems. The resulting effects can range from noise on communication lines and errors in digital circuits to equipment damage and even personnel hazards.

Some of the more well-known sources of electromagnetic fields include nearby lightning strikes; AM, TV, and FM broadcast stations; radars; industrial, scientific, and medical (ISM) equipment; automobile ignitions; personnel electrostatic discharge; the esoteric nuclear electromagnetic pulse (NEMP); and power supply noise and switching transients inside electronic equipment.

The list of susceptible conductors is as diverse as the list of sources and spans the range from long CATV cable runs, overhead power lines, and telephone lines to short interconnecting wires and cables inside electronic products. Regardless of the nature of the source or the type of conductor, the problem is the same: given the field, find the induced current.

This book is intended for those who must deal quantitatively with the coupling of electromagnetic fields to transmission lines, whether the particular application happens to be in the field of communication, data processing, power, telephone, or control systems. The systems engineer and circuit designer interested in numerical solutions will find abundant application data in the form of solved examples and spectrum profiles. The electromagnetic specialist will find sufficient theoretical detail to permit the extension of the results to a wide variety of problems. Academicians may find this treatment, which applies both transmission line theory and electromagnetic theory to a practical problem of growing interest, an ideal vehicle for a graduate or advanced undergraduate course.

The material in this book is derived from the author's research on the coupling of nonuniform electromagnetic fields to transmission lines and subsequent application of the theoretical results to communication problems.

Acknowledgement of the work of previous investigators is made in the form of a chapter-by-chapter bibliography at the end of the book.

I would like to thank Dr. Arvind Shah, IBM Research Triangle Park, N.C., for the data on cable transfer impedance found in Chapter 4. The encouragement and assistance of Joseph Weglarz and Ron Freeman, IBM, Kingston, New York, is greatly appreciated.

ALBERT A. SMITH, JR.

Woodstock, New York
August 1976

CONTENTS

COUPLING OF EXTERNAL
ELECTROMAGNETIC FIELDS
TO TRANSMISSION LINES

ONE

COUPLING MODEL

$\mathbf{T}$his chapter develops the theory of excitation of a two-wire transmission line illuminated by an external electromagnetic field and provides the mathematical foundation for succeeding chapters. Equations for the differential mode current flowing along the line and in the terminations are derived. A method to calculate the common mode current when the line is near a ground plane is described. Solutions for a wire over a ground plane are developed by analogy with the two-wire line using the method of images.

1.1 TWO-WIRE LINE

An isolated two-wire transmission line illuminated by a nonuniform electromagnetic field is shown in Fig. 1-1. The line is contained in the x–z plane, with the conductors parallel to the z axis and the terminations parallel to the x axis. The length of the line is s. The wires are spaced a distance b apart and have a diameter a. Z_1 and Z_2 are, respectively, the left-hand and right-hand terminating impedances.

$E(x,y,z)$ is the electric field component of the incident wave, and $H(x,y,z)$ is the magnetic field component. The response of the line can be found entirely in terms of the E_x field along the terminations and the E_z field along the conductors. The solution is obtained by integrating the responses to the infinitesimal voltage sources produced by the field along the terminations and along the conductors (see Fig. 1-2). The response of the line could just as well be given in terms of the magnetic field alone. However, the resulting equations involve the space derivatives of all three magnetic field components and are more cumbersome than those obtained from the electric field formulation.

By convention, the infinitesimal voltage generators along the line due to E_z are separated into their differential mode (DM) and common mode (CM) components. This is illustrated in Fig. 1-3, which also shows the differential and common mode currents in the conductors. (The term differential mode is also variously referred to in the literature as transmis-

2

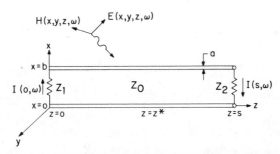

FIGURE 1-1. Two-wire line illuminated by a nonuniform electromagnetic field.

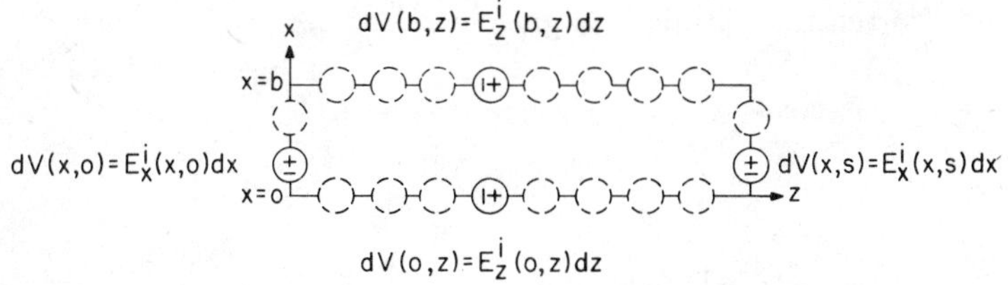

$$x \quad dV(b,z) = E_z^i(b,z)\,dz$$

$$dV(x,0) = E_x^i(x,0)\,dx$$

$$dV(x,s) = E_x^i(x,s)\,dx$$

$$dV(0,z) = E_z^i(0,z)\,dz$$

FIGURE 1-2. Infinitesimal voltage generators along the conductors and along the terminations.

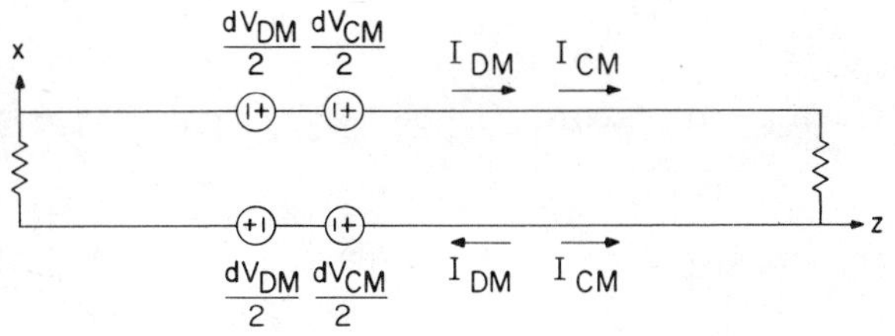

$$\frac{dV_{DM}}{2} \quad \frac{dV_{CM}}{2} \qquad I_{DM} \quad I_{CM}$$

$$\frac{dV_{DM}}{2} \quad \frac{dV_{CM}}{2} \qquad I_{DM} \quad I_{CM}$$

$$dV_{DM}(z) = dV(b,z) - dV(0,z) = \left[E_z^i(b,z) - E_z^i(0,z) \right] dz$$

$$dV_{CM}(z) = dV(b,z) + dV(0,z) = \left[E_z^i(b,z) + E_z^i(0,z) \right] dz$$

FIGURE 1-3. Symmetrical and antisymmetrical parts of the generators along the conductors due to E_z.

sion line mode and antisymmetrical, bidirectional, and odd mode. The term common mode is synonymous with antenna mode, dipole mode, and symmetrical, codirectional, and even mode).

Only the differential mode current flows in the terminating impedances and thus, in most applications, it is the only current of interest. The common mode current is zero at the ends of the line. Moving away from the terminations, the common mode current increases rapidly to values much greater than the differential mode current. In general, the current in each conductor at any point along the line is the sum of the differential mode and common mode components.

1.2 DIFFERENTIAL MODE CURRENT

The differential mode current on the line in Fig. 1-1 is found using straightforward transmission line theory, which applies when the conductor spacing is much less than a wavelength. For a dissipative line, the differential mode current at any point z^* along the line (i.e., the differential mode current distribution) is given by

$$I(z^*,\omega) = \frac{Z_0 \cosh \gamma(s-z^*) + Z_2 \sinh \gamma(s-z^*)}{Z_0 D}$$

$$\times \int_0^{z^*} K(z,\omega) \left[Z_0 \cosh \gamma z + Z_1 \sinh \gamma z \right] dz$$

$$+ \frac{Z_0 \cosh \gamma z^* + Z_1 \sinh \gamma z^*}{Z_0 D}$$

$$\times \int_{z^*}^{s} K(z,\omega) \left[Z_0 \cosh \gamma(s-z) + Z_2 \sinh \gamma(s-z) \right] dz$$

$$+ \frac{1}{D} \left[Z_0 \cosh \gamma(s-z^*) + Z_2 \sinh \gamma(s-z^*) \right] \int_0^b E_x^i(x,0,\omega) dx$$

$$- \frac{1}{D} \left[Z_0 \cosh \gamma z^* + Z_1 \sinh \gamma z^* \right] \int_0^b E_x^i(x,s,\omega) dx \qquad (1\text{-}1)$$

where

$Z_0 = \sqrt{Z/Y}$ characteristic impedance
Z distributed series impedance of line
Y distributed shunt admittance of line
$K(z,\omega) = E_z^i(b,z,\omega) - E_z^i(0,z,\omega)$
$D = (Z_0 Z_1 + Z_0 Z_2) \cosh \gamma s + (Z_0^2 + Z_1 Z_2) \sinh \gamma s$
$E_z^i(b,z,\omega)$ field in z direction incident on upper conductor ($x=b$)
$E_z^i(0,z,\omega)$ field in z direction incident on lower conductor ($x=0$)

$E_x^i(x,0,\omega)$ field in x direction incident on left-hand termination ($z=0$)
$E_x^i(x,s,\omega)$ field in x direction incident on right-hand termination ($z=s$)
$\gamma=\alpha+j\beta$ propagation constant of line
α attenuation constant of line
$\beta=2\pi/\lambda$ phase constant of line
$\lambda=$ wavelength
$\omega=2\pi f$
$f=$ frequency, hertz.

The load currents are obtained by setting $z^*=0$ and $z^*=s$ in (1-1). The current in the left-hand and right-hand terminations are, respectively,

$$I(0,\omega)=\frac{1}{D}\int_0^s K(z,\omega)\big[Z_0\cosh\gamma(s-z)+Z_2\sinh\gamma(s-z)\big]\,dz$$

$$+\frac{1}{D}\big[Z_0\cosh\gamma s+Z_2\sinh\gamma s\big]\int_0^b E_x^i(x,0,\omega)\,dx$$

$$-\frac{Z_0}{D}\int_0^b E_x^i(x,s,\omega)\,dx \tag{1-2}$$

and

$$I(s,\omega)=\frac{1}{D}\int_0^s K(z,\omega)\big[Z_0\cosh\gamma z+Z_1\sinh\gamma z\big]\,dz$$

$$+\frac{Z_0}{D}\int_0^b E_x^i(x,0,\omega)\,dx$$

$$-\frac{1}{D}\big[Z_0\cosh\gamma s+Z_1\sinh\gamma s\big]\int_0^b E_x^i(x,s,\omega)\,dx. \tag{1-3}$$

The voltages across the terminations are

$$V(0,\omega)=I(0,\omega)Z_1$$

$$V(s,\omega)=I(s,\omega)Z_2.$$

In most practical instances, the attenuation on transmission lines is negligible, that is, $\alpha\doteq0$. The differential mode current distribution on a lossless line, substituting $\gamma = j\beta$ in (1-1), is

$$I(z^*,\omega) = \frac{Z_0 \cos\beta(s-z^*)+jZ_2\sin\beta(s-z^*)}{Z_0 D}$$

$$\times \int_0^{z^*} K(z,\omega)\left[Z_0\cos\beta z+jZ_1\sin\beta z\right]dz$$

$$+\frac{Z_0\cos\beta z^*+jZ_1\sin\beta z^*}{Z_0 D}$$

$$\times \int_{z^*}^s K(z,\omega)\left[Z_0\cos\beta(s-z)+jZ_2\sin\beta(s-z)\right]dz$$

$$+\frac{1}{D}\left[Z_0\cos\beta(s-z^*)+jZ_2\sin\beta(s-z^*)\right]\int_0^b E_x^i(x,0,\omega)\,dx$$

$$-\frac{1}{D}\left[Z_0\cos\beta z^*+jZ_1\sin\beta z^*\right]\int_0^b E_x^i(x,s,\omega)\,dx \qquad (1\text{-}4)$$

where now

$$D=(Z_0 Z_1 + Z_0 Z_2)\cos\beta s + j(Z_0^2 + Z_1 Z_2)\sin\beta s$$
$$Z_0 = 120\ln(2b/a)=276\log_{10}(2b/a) \text{ characteristic impedance.}^\dagger$$

The load currents for a lossless line are given by

$$I(0,\omega)=\frac{1}{D}\int_0^s K(z,\omega)\left[Z_0\cos\beta(s-z)+jZ_2\sin\beta(s-z)\right]dz$$

$$+\frac{1}{D}\left[Z_0\cos\beta s+jZ_2\sin\beta s\right]\int_0^b E_x^i(x,0,\omega)\,dx$$

$$-\frac{Z_0}{D}\int_0^b E_x^i(x,s,\omega)\,dx \qquad (1\text{-}5)$$

and

$$I(s,\omega)=\frac{1}{D}\int_0^s K(z,\omega)\left[Z_0\cos\beta z+jZ_1\sin\beta z\right]dz+\frac{Z_0}{D}\int_0^b E_x^i(x,0,\omega)\,dx$$

$$-\frac{1}{D}\left[Z_0\cos\beta s+jZ_1\sin\beta s\right]\int_0^b E_x^i(x,s,\omega)\,dx \qquad (1\text{-}6)$$

where, again,
$$D=(Z_0 Z_1 + Z_0 Z_2)\cos\beta s + j(Z_0^2 + Z_1 Z_2)\sin\beta s$$
$$Z_0 = 120\ln(2b/a).$$

---

†For convenience, a graph of the characteristic impedance of a lossless two-conductor line is given in Appendix A.

1.3 COMMON MODE CURRENT BY THE METHOD OF IMAGES

The common mode or antenna mode current along an isolated two-wire transmission line must be obtained by the methods of linear antenna theory. The antenna theory approach, which is not treated here, requires the solution of complicated integral equations, even for relatively simple illuminating fields. Fortunately, most transmission lines are not isolated; that is, they are parallel, or approximately parallel, to a conducting ground plane or to earth. When this is the case, the common mode current distribution along the line can be obtained from transmission line theory by treating the line and its image in the ground plane as a two-wire transmission line open circuited at both ends. The equations are given in Section 1.4.

When applying this method, which is also useful for multiconductor transmission lines, the conductor spacing must be negligible compared with the distance to the ground plane. All the conductors in the cable are treated as a single wire with a diameter equal to, approximately, the overall diameter of the cable. (The exact equivalent diameter is not critical since it is used only in the calculation of the characteristic impedance of the line, which varies relatively slowly as a function of wire diameter). The common mode current which is calculated is then assumed to divide equally among the conductors in the cable.

The method of images is valid as long as the transmission line is not too far from the ground plane in terms of the line length. For spacings between the transmission line and ground of one-fourth the line length or less, the difference between the common mode currents calculated by antenna theory and by the method of images is less than 2 dB. Whitescarver [3], for instance, obtained excellent results using the method of images to predict the antenna mode current on a 40-ft-long transmission line 6 ft above the surface of the earth, assuming the earth to be a perfect ground (infinite conductivity).

1.4 WIRE OVER A GROUND PLANE

A single-wire transmission line terminated at both ends to a perfectly conducting ground plane is shown in Fig. 1-4a. The distance between the wire and the ground plane is $b/2$, and the terminating impedances are $Z_1/2$ and $Z_2/2$. The length of the line is s, the diameter of the wire is a, and E^i is the incident electric field.

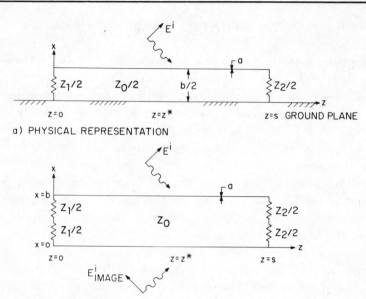

a) PHYSICAL REPRESENTATION

b) IMAGE PROBLEM

FIGURE 1-4. Wire over a ground plane.

Figure 1-4b shows the ground plane replaced by the image of the wire and its terminations. The necessary image field is also shown. The image and direct fields satisfy the appropriate boundary conditions at the position of the ground plane, that is, $E_z = 0$ and $E_x = 2E_x^i$. Except for the image field, this is the exact problem solved in the previous sections.

When both the incident and image field excitations are taken into account, two things are apparent. First, the differential mode driving function for the wire over a ground is twice that for an isolated two-wire line. Second, the common mode driving function is zero, that is, the common mode or antenna mode current does not exist on the wire over a ground plane. This is illustrated in Fig. 1-5 for the E_z field. Figure 1-5a shows the infinitesimal voltage generators at one point on the line induced by the incident and image fields. By symmetry

$$E_{z_{\text{image}}}^i(b,z) = -E_z^i(0,z)$$

and

$$E_{z_{\text{image}}}^i(0,z) = -E_z^i(b,z).$$

Making this substitution, the result is shown in Fig. 1-5b, where it is evident that the differential mode driving function is twice that shown in Fig. 1-3 for an isolated two-wire line. The common mode driving function is zero.

The equations for the wire over a ground plane shown in Fig. 1-4 can be obtained by analogy from equations (1-1) to (1-6) using excitation functions that are greater by a factor of 2. For a dissipative wire over a ground plane, the current distribution is

$$I(z^*,\omega) = \frac{Z_0\cosh\gamma(s-z^*)+Z_2\sinh\gamma(s-z^*)}{Z_0 D}$$

$$\times \int_0^{z^*} 2K(z,\omega)\left[Z_0\cosh\gamma z + Z_1\sinh\gamma z\right]dz$$

$$+ \frac{Z_0\cosh\gamma z^* + Z_1\sinh\gamma z^*}{Z_0 D} \int_{z^*}^{s} 2K(z,\omega)\left[Z_0\cosh\gamma(s-z)\right.$$

$$\left.+ Z_2\sinh\gamma(s-z)\right]dz + \frac{2}{D}\left[Z_0\cosh\gamma(s-z^*)\right.$$

$$\left.+ Z_2\sinh\gamma(s-z^*)\right]\int_0^b E_x^i(x,0,\omega)dx$$

$$- \frac{2}{D}\left[Z_0\cosh\gamma z^* + Z_1\sinh\gamma z^*\right]\int_0^b E_x^i(x,s,\omega)dx \qquad (1-7)$$

where now

$Z_0 = \sqrt{Z/Y}$ characteristic impedance of equivalent two-wire line (i.e., the wire and its image)

Z distributed series impedance of equivalent two-wire line

Y distributed shunt admittance of equivalent two-wire line

$K(z,\omega) = E_z^i(b,z,\omega) - E_z^i(0,z,\omega)$

$D = (Z_0 Z_1 + Z_0 Z_2)\cosh\gamma s + (Z_0^2 + Z_1 Z_2)\sinh\gamma s$

$E_z^i(b,z,\omega)$ incident field in z direction on wire ($x=b$)

$E_z^i(0,z,\omega)$ incident field in z direction on image conductor ($x=0$)

$E_x^i(x,0,\omega)$ incident field in x direction on left-hand termination ($z=0$)

$E_x^i(x,s,\omega)$ incident field in x direction on right-hand termination ($z=s$)

$\gamma = \alpha + j\beta$ propagation constant of equivalent two-wire line

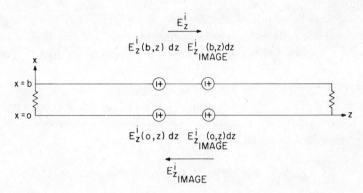

a) INFINITESIMAL GENERATORS DUE TO INCIDENT AND IMAGE FIELDS.

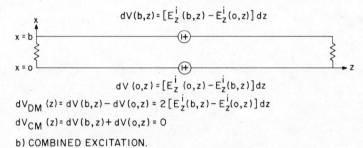

b) COMBINED EXCITATION.

FIGURE 1-5. Excitation functions for a wire over a ground plane due to E_z.

α attenuation constant of equivalent two-wire line

$\beta = 2\pi/\lambda$ phase constant

λ wavelength

$\omega = 2\pi f$

f frequency, hertz.

The load currents in the left-hand and right-hand terminations are, respectively,

$$I(0,\omega) = \frac{2}{D}\int_0^s K(z,\omega)\Big[Z_0\cosh\gamma(s-z) + Z_2\sinh\gamma(s-z)\Big]dz$$

$$+ \frac{2}{D}\Big[Z_0\cosh\gamma s + Z_2\sinh\gamma s\Big]\int_0^b E_x^i(x,0,\omega)dx$$

$$- \frac{2Z_0}{D}\int_0^b E_x^i(x,s,\omega)dx \tag{1-8}$$

and

$$I(s,\omega) = \frac{2}{D} \int_0^s K(z,\omega)\left[Z_0 \cosh \gamma z + Z_1 \sinh \gamma z \right] dz$$

$$+ \frac{2Z_0}{D} \int_0^b E_x^i(x,0,\omega)\,dx$$

$$- \frac{2}{D}\left[Z_0 \cosh \gamma s + Z_1 \sinh \gamma s \right] \int_0^b E_x^i(x,s,\omega)\,dx. \qquad (1\text{-}9)$$

The voltages across the actual terminations are

$$V(0,\omega) = I(0,\omega) \times (Z_1/2)$$

$$V(s,\omega) = I(s,\omega) \times (Z_2/2).$$

For a lossless wire over a ground plane, the current distribution is given by

$$I(z^*,\omega) = \frac{Z_0 \cos \beta (s - z^*) + jZ_2 \sin \beta (s - z^*)}{Z_0 D}$$

$$\times \int_0^{z^*} 2K(z,\omega)\left[Z_0 \cos \beta z + jZ_1 \sin \beta z \right] dz$$

$$+ \frac{Z_0 \cos \beta z^* + jZ_1 \sin \beta z^*}{Z_0 D}$$

$$\times \int_{z^*}^s 2K(z,\omega)\left[Z_0 \cos \beta (s - z) + jZ_2 \sin \beta (s - z) \right] dz$$

$$+ \frac{2}{D}\left[Z_0 \cos \beta (s - z^*) + jZ_2 \sin \beta (s - z^*) \right]$$

$$\times \int_0^b E_x^i(x,0,\omega)\,dx - \frac{2}{D}\left[Z_0 \cos \beta z^* + jZ_1 \sin \beta z^* \right]$$

$$\times \int_0^b E_x^i(x,s,\omega)\,dx \qquad (1\text{-}10)$$

where now

$$D = (Z_0 Z_1 + Z_0 Z_2)\cos \beta s + j(Z_0^2 + Z_1 Z_2)\sin \beta s$$
$$Z_0 = 276 \log_{10}(2b/a) \text{ characteristic impedance of equivalent two-wire line.}$$

The load currents for a lossless wire over a ground plane are

$$I(0,\omega) = \frac{2}{D} \int_0^s K(z,\omega) \left[Z_0 \cos\beta\,(s-z) + jZ_2 \sin\beta\,(s-z) \right] dz$$

$$+ \frac{2}{D} \left[Z_0 \cos\beta s + jZ_2 \sin\beta s \right] \int_0^b E_x^i(x,0,\omega)\,dx$$

$$- \frac{2Z_0}{D} \int_0^b E_x^i(x,s,\omega)\,dx \qquad (1\text{-}11)$$

and

$$I(s,\omega) = \frac{2}{D} \int_0^s K(z,\omega) \left[Z_0 \cos\beta z + jZ_1 \sin\beta z \right] dz$$

$$+ \frac{2Z_0}{D} \int_0^b E_x^i(x,0,\omega)\,dx$$

$$- \frac{2}{D} \left[Z_0 \cos\beta s + jZ_1 \sin\beta s \right] \int_0^b E_x^i(x,s,\omega)\,dx \qquad (1\text{-}12)$$

where D and Z_0 are as defined in equation (1-10).

As discussed in Section 1.3, the method of images used to solve the wire over a ground plane problem can be used to find the antenna mode current distribution on a two-wire transmission line run parallel to (but not connected to) a conducting ground plane or earth. The solution is obtained directly from equation (1-10) by setting $Z_1 = Z_2 = \infty$. The result is

$$I(z^*,\omega) = \frac{j2\sin\beta\,(s-z^*) \int_0^z K(z,\omega)\sin\beta z\,dz}{Z_0 \sin\beta s}$$

$$+ \frac{j2\sin\beta z^* \int_{z^*}^s K(z,\omega)\sin\beta\,(s-z)\,dz}{Z_0 \sin\beta s}. \qquad (1\text{-}13)$$

1.5   TIME DOMAIN SOLUTIONS

Following the usual convention, the solutions in this book are in the frequency domain and are of the form

$$I(\omega) = E(\omega)H(\omega)$$

where $I(\omega)$ is the current spectrum, $E(\omega)$ is the spectrum of the illuminating field, and $H(\omega)$ is a function of the transmission line parameters and terminating impedances.

In theory at least, the time domain response can be obtained by applying the inverse Fourier transform. However, the complexity of the formulas for $H(\omega)$ and the difficulty of obtaining a complete description of the electric field spectrum (magnitude and phase) makes this process practical in only a limited number of situations.

If the time history $E(t)$ of the incident field is known, the frequency spectrum obtained from the Fourier transform is

$$E(\omega) = \int_{-\infty}^{\infty} E(t)\epsilon^{-j\omega t}\, dt.$$

Then, the time domain response for the current is

$$I(t) = \frac{1}{2\pi} \int_{-\infty}^{\infty} E(\omega)H(\omega)\epsilon^{j\omega t}\, d\omega.$$

A practical time domain solution for plane wave excitation of a lossless line is given in Section 2.4.

1.6 A NOTE ON THE DENOMINATOR FUNCTION D

The denominator function D for a lossless line is given by

$$D = (Z_0 Z_1 + Z_0 Z_2)\cos\beta s + j(Z_0^2 + Z_1 Z_2)\sin\beta s.$$

To avoid poles in the solutions for the current due to zeros in the denominator function, it is necessary only to match Z_1 or Z_2 (or both) to the characteristic impedance Z_0. For example, when $Z_1 = Z_0$,

$$D = Z_0^2 \left(1 + \frac{Z_2}{Z_0}\right)\epsilon^{-j\beta s} \neq 0$$

and when Z_2 is real,

$$|D| = Z_0^2 \left(1 + \frac{Z_2}{Z_0}\right) = \text{constant.}$$

When both ends of the line are matched, $Z_1 = Z_2 = Z_0$ and

$$D = 2Z_0^2 \epsilon^{-j\beta s}$$

and

$$|D| = 2Z_0^2.$$

When the line is shorted at both ends, $Z_1 = Z_2 = 0$ and

$$D = jZ_0^2 \sin \beta s.$$

In this case, poles in the solution occur (unless cancelled out by the numerator) at

$$f_{\text{MHz}} = \frac{150n}{s} \qquad n = 0, 1, 2, \ldots$$

where f_{MHz} is the frequency in megahertz.

1.7 DISSIPATIVE LOSSES

Attenuation due to dissipative losses in the conductors and in the dielectric medium can be completely neglected in the analytical solution of many practical transmission line problems. Fortunately this is so, because the resulting transmission line equations are considerably simplified compared to the general lossy line case. In this section we briefly examine the conditions (line length, frequency, conductor size and spacing, and the material of the conductors and the dielectric medium) under which the lossless assumption is justified for radio frequency lines.

The characteristic impedance Z_0 of a uniform transmission line is

$$Z_0 = \sqrt{\frac{R + j\omega L}{G + j\omega C}} = \sqrt{\frac{L}{C}} \sqrt{\frac{1 + R/j\omega L}{1 + G/j\omega C}} \qquad (1\text{-}14)$$

where R, G, L, and C are, respectively, the resistance, conductance, inductance, and capacitance per unit length of the line.

At the low end of the spectrum, the characteristic impedance is a frequency-dependent complex number. The frequency range where Z_0 is complex depends on the particular line constants. For typical open-wire telephone lines, the range includes voice frequencies and below. At zero frequency, of course, $Z_0 = (R/G)^{1/2}$.

At the higher frequencies, $\omega L \gg R$ and $\omega C \gg G$, and the characteristic impedance is given by

$$Z_0 = \sqrt{\frac{L}{C}} \qquad (1\text{-}15)$$

which is real-valued and independent of frequency.

The propagation constant γ of a uniform transmission line is

$$\gamma = \alpha + j\beta = \sqrt{(R + j\omega L)(G + j\omega C)} \qquad (1\text{-}16)$$

or

$$\gamma = j\omega\sqrt{LC}\left(1 + \frac{R}{j\omega L}\right)^{1/2}\left(1 + \frac{G}{j\omega C}\right)^{1/2}. \qquad (1\text{-}17)$$

At the higher frequencies where $\omega L \gg R$ and $\omega C \gg G$, the last two terms of (1-17) may be expanded using the binomial theorem $(1 + X)^n = 1 + nX + \cdots$, retaining the first two terms only.
Then

$$\gamma = j\omega\sqrt{LC}\left(1 + \frac{1}{2}\frac{R}{j\omega L}\right)\left(1 + \frac{1}{2}\frac{G}{j\omega C}\right).$$

Performing the multiplication and neglecting the term containing the product RG, we have

$$\gamma = \frac{R}{2}\sqrt{\frac{C}{L}} + \frac{G}{2}\sqrt{\frac{L}{C}} + j\omega\sqrt{LC}$$

or, using the high-frequency characteristic impedance in (1-15),

$$\gamma = \frac{R}{2Z_0} + \frac{GZ_0}{2} + j\omega\sqrt{LC}. \qquad (1\text{-}18)$$

The attenuation and phase constants are

$$\alpha = \frac{R}{2Z_0} + \frac{GZ_0}{2} \qquad \text{nepers/meter}$$

$$\beta = \omega\sqrt{LC} \qquad \text{radians/meter}.$$

On most transmission lines, and especially on air-dielectric lines, the dissipative losses due to the shunt conductance G are negligible compared to the resistive losses in the conductors, and the attenuation constant can be taken as

$$\alpha = \frac{R}{2Z_0} \qquad \text{nepers/meter}. \qquad (1\text{-}19)$$

The resistance per unit length R of a single solid cylindrical conductor of diameter a is equal to the D.C. resistance of a fictitious hollow cylindrical conductor with the same outside diameter and with a wall thickness equal to the depth of penetration. The D.C. resistance is just the resistivity ρ of the metal divided by the cross-sectional area. Then

$$R = \frac{\rho}{\pi a \delta} = \frac{1}{a}\sqrt{\frac{\rho f \mu}{\pi}}$$

where

$$\delta = \sqrt{\frac{\rho}{\pi f \mu}} \qquad \text{depth of penetration.}$$

The resistance per unit length of a two-wire transmission line is, then

$$R = \frac{2}{a}\sqrt{\frac{\rho f \mu}{\pi}} \qquad \text{ohms/meter,} \qquad (1\text{-}20)$$

which may be expressed in terms of the relative resistivity ρ_r and relative permeability μ_r as

$$R = 1.66 \times 10^{-7}\frac{\sqrt{\rho_r \mu_r f}}{a} \qquad \text{ohms/meter} \qquad (1\text{-}21)$$

where

$\rho = \rho_c \rho_r$
ρ_c resistivity of copper, 1.72×10^{-8} ohm-meters
ρ_r relative resistivity (referenced to copper)
$\mu = \mu_0 \mu_r$
μ_0 permeability of air, $4\pi \times 10^{-7}$ henrys/meter
μ_r relative permeability
f frequency, hertz
a conductor diameter, meters.

Table 1-1 gives the relative resistivity and relative permeability of a number of metals.

TABLE 1-1. Relative Resistivities
and Permeabilities

Metal	ρ_r	μ_r
Silver	0.95	1
Copper	1	1
Aluminum	1.64	1
Brass	3.9	1
Cast Iron	5.6	60
Tin	6.7	1
Steel	7.6–12.7	300

The attenuation constant in equation (1-19), on substitution of (1-21), becomes

$$\alpha = 8.3 \times 10^{-8} \frac{\sqrt{\rho_r \mu_r f}}{a Z_0} .$$

For two-conductor air-dielectric lines

$$Z_0 = 276 \log_{10} \frac{2b}{a} ,$$

and the attenuation constant is given by

$$\alpha = 3 \times 10^{-10} \frac{\sqrt{\rho_r \mu_r f}}{a \log_{10} 2b/a} \qquad \text{nepers/meter} \qquad (1\text{-}22)$$

or

$$\alpha = 2.6 \times 10^{-9} \frac{\sqrt{\rho_r \mu_r f}}{a \log_{10} 2b/a} \qquad \text{decibels/meter}$$

where the conductor spacing b and conductor diameter a are in meters, and the frequency f is in hertz.

Equation (1-22) is plotted in Fig. 1-6 for copper, aluminum, and steel transmission lines with a conductor spacing of 10 centimeters and a wire diameter of 1 millimeter. Figure 1-7 shows the attenuation constant for lines of the same materials, but with a conductor spacing of 1 meter and a

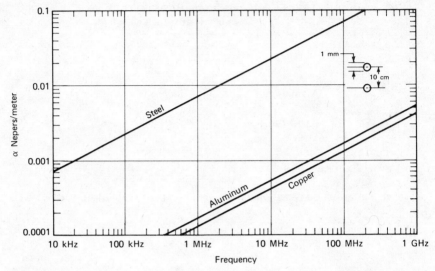

FIGURE 1-6. Attenuation constant of two-wire air-dielectric line ($b = 10$ cm, $a = 1$ mm).

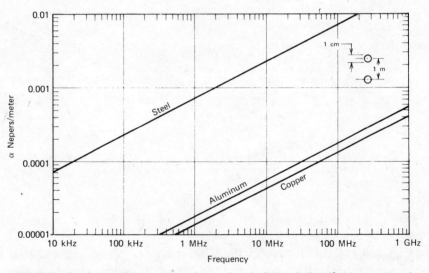

FIGURE 1-7. Attenuation constant of two-wire air-dielectric line ($b = 1$ m, $a = 1$ cm).

wire diameter of 1 centimeter. (The attenuation constant in decibels/meter can be found by multiplying the ordinates in nepers/meter by 8.686).

The low attenuation constants for the copper and aluminum lines in Figs. 1-6 and 1-7 illustrates that the lossless assumption is a valid one, even at the higher frequencies, for many practical length lines composed of these materials. For example, in Fig. 1-6 ($a = 1$ mm, $b = 10$ cm), the attenuation constant is 0.005 nepers/meter (0.0434 dB/m) at 1 GHz. For a line 10 m long, the attenuation experienced by a wave traveling the length of the line is only 0.05 nepers (0.434 dB) at 1 GHz.

For the steel conductor line in Fig. 1-6, the attenuation constant is 0.07 nepers/meter (0.6 dB/m) at 100 MHz, which is negligible only on very short lines.

In practice, mismatch losses, which arise from series and shunt discontinuities along the line, can result in greater attenuation to signals traveling on the line than dissipative losses. The mismatch losses result from scattering or reflection of the waves at the discontinuities.

The effect of dissipative losses on the load current spectrum of a line illuminated by a plane wave is analyzed in Section 2.5.

CHAPTER
TWO

PLANE-WAVE EXCITATION

$\mathbf{T}$he general equations in Chapter 1 reduce to relatively simple forms when the incident field is a plane wave. In this chapter formulas are derived for the load currents of lossless transmission lines illuminated by plane waves parallel to the terminations and parallel to the conductors. Numerical examples are also given.

2.1 ELECTRIC FIELD PARALLEL TO THE TERMINATIONS

Figure 2-1 illustrates a plane wave with the electric field parallel to the terminations incident on the line at an angle Φ. Both H^i and the direction of propagation are in the y–z plane.

Since $E_z^i = 0$, $K(z) = 0$. Also, E_x^i is constant (uniform) over the length of the terminations, that is,

$$\int_0^b E_x^i \, dx = bE_x^i.$$

Then the load current in the right-hand termination, from equation (1-6), is just

$$I(s,\omega) = \frac{bZ_0 E_x^i(0,\omega) - bE_x^i(s,\omega)\left[Z_0 \cos \beta s + jZ_1 \sin \beta s\right]}{D}. \qquad (2\text{-}1)$$

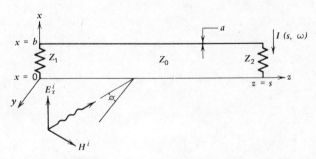

FIGURE 2-1. E_x incident at angle Φ.

22

If the phase reference for the incoming field is taken at $z = 0$, we have

$$E_x^i(0, \omega) = E_x^i$$

$$E_x^i(s, \omega) = E_x^i \varepsilon^{-j\beta s \sin \Phi}.$$

Substituting these relations into (2-1) and performing the algebra results in

$$I(s, \omega) = \frac{bE_x^i(\omega)}{2D} \left\{ 2Z_0 - (Z_0 - Z_1) \cos \left[\beta s(1 + \sin \Phi) \right] \right.$$

$$- (Z_0 + Z_1) \cos \left[\beta s(1 - \sin \Phi) \right]$$

$$+ j(Z_0 - Z_1) \sin \left[\beta s(1 + \sin \Phi) \right]$$

$$\left. - j(Z_0 + Z_1) \sin \left[\beta s(1 - \sin \Phi) \right] \right\}. \qquad (2\text{-}2)$$

For a plane wave traveling in the z direction as shown in Fig. 2-2, $\Phi = 90°$ and (2-2) reduces to

$$I(s, \omega) = \frac{bE_x^i(\omega)}{2D}(Z_0 - Z_1)\left[(1 - \cos 2\beta s) + j \sin 2\beta s \right]. \qquad (2\text{-}3)$$

The load current spectrum normalized to the electric field spectrum is

$$\frac{I(s, \omega)}{E_x^i(\omega)} = \frac{b(Z_0 - Z_1)}{2D}\left[(1 - \cos 2\beta s) + j \sin 2\beta s \right]. \qquad (2\text{-}4)$$

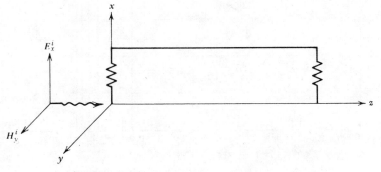

FIGURE 2-2. E_x traveling in the z direction ($\Phi = 90°$).

The magnitude of equation (2-4) in decibels, that is,

$$20\log_{10}\left|\frac{I(s,\omega)}{E_x^i(\omega)}\right| \qquad (2\text{-}5)$$

is plotted in Figs. 2-3 and 2-4 for two different line geometries. In each figure the results for three different combinations of load impedances are given.

To satisfy a wide diversity of applications, numerical results are given throughout this book for both long and short lines wherever practical. For example, the dimensions of the 10-m-long line in Fig. 2-3 is representative of external cables, power lines, telephone lines, metal pipes, and the like. The 1-m line in Figure 2-4 has dimensions representative of wire and cables found inside electronic equipment. Although it is impossible to give numerical results for all possible situations, these two line length examples at least illustrate the effects of the major line parameters such as length and conductor spacing on the currents excited on the line.

Figure 2-5 shows a plane wave traveling in the $-y$ direction. In this

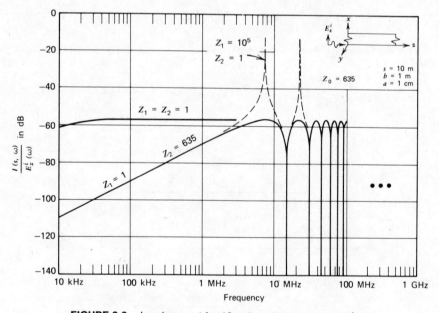

FIGURE 2-3. Load current for 10-m-long line excited by $E_x^i(z)$.

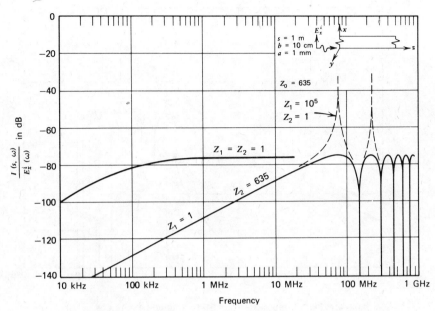

FIGURE 2-4. Load current for 1-m-long line excited by $E_x^i(z)$.

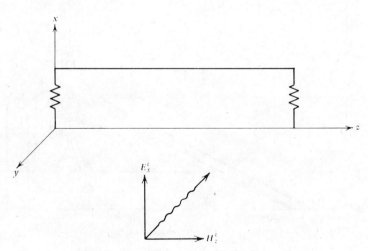

FIGURE 2-5. E_x traveling in the $-y$ direction ($\Phi = 0°$).

25

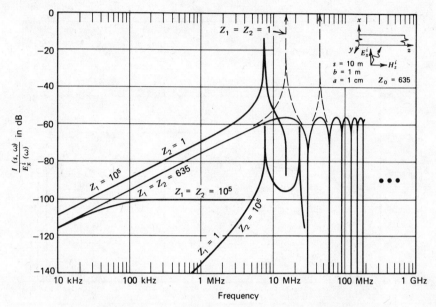

FIGURE 2-6. Load current for 10-m-long line excited by $E_x^i(y)$.

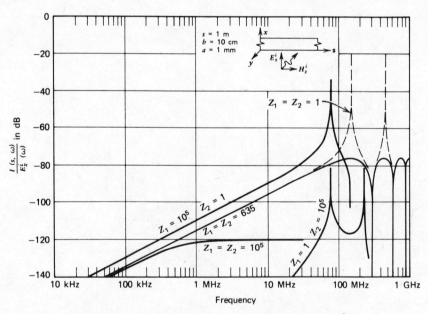

FIGURE 2-7. Load current for 1-m-long line excited by $E_x^i(y)$.

26

case, $\Phi = 0°$ and (2-2) reduces to

$$\frac{I(s,\omega)}{E_x^i(\omega)} = \frac{b}{D}\left[Z_0(1-\cos\beta s) - jZ_1\sin\beta s\right]. \qquad (2\text{-}6)$$

The magnitude of (2-6) is plotted in Figs. 2-6 and 2-7 for two different line geometries and five combinations of load impedances.

2.2   ELECTRIC FIELD PARALLEL TO THE CONDUCTORS

A plane wave arriving at an angle Ψ and polarized such that the electric field vector is parallel to the transmission line conductors is shown in Fig. 2-8. The magnetic field vector and direction of propagation are in the x–y plane.

Since $E_x^i = 0$ and E_z^i is uniform over each conductor (i.e., E_z^i is independent of z), the load current given by equation (1-6) reduces to

$$I(s,\omega) = \frac{K(\omega)}{D}\int_0^s\left[Z_0\cos\beta z + jZ_1\sin\beta z\right]dz$$

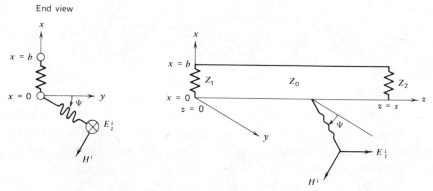

FIGURE 2-8. E_z incident at angle Ψ.

or

$$I(s,\omega) = \frac{K(\omega)}{\beta D}\left[Z_0\sin\beta s + jZ_1(1-\cos\beta s)\right] \qquad (2\text{-}7)$$

where

$$K(\omega) = E_z^i(b,\omega) - E_z^i(0,\omega).$$

If the phase reference is taken midway between the conductors at $x = b/2$,

$$E_z^i(b) = E_z^i \varepsilon^{-j(\beta b/2)\sin\Psi}$$

$$E_z^i(0) = E_z^i \varepsilon^{+j(\beta b/2)\sin\Psi}.$$

Then,

$$K(\omega) = E_z^i(\omega)\left[\varepsilon^{-j(\beta b/2)\sin\Psi} - \varepsilon^{+j(\beta b/2)\sin\Psi}\right].$$

Using the identity $\varepsilon^{\pm jw} = \cos w \pm j\sin w$,

$$K(\omega) = E_z^i(\omega)\left[-j2\sin\left(\frac{\beta b}{2}\sin\Psi\right)\right]. \tag{2-8}$$

The final expression for the load current, obtained by substituting (2-8) into (2-7), is

$$I(s,\omega) = \frac{E_z^i(\omega)\left[-j2\sin\left(\dfrac{\beta b}{2}\sin\Psi\right)\right]}{\beta D}$$

$$\times \left[Z_0\sin\beta s + jZ_1(1 - \cos\beta s)\right]. \tag{2-9}$$

When $\Psi = 0$, (2-9) yields $I(s,\omega) = 0$. This agrees with the "physical" picture. The differential mode current on the line is the difference between the currents excited on the two conductors. Since the electric field arrives at each conductor simultaneously, the currents excited on the two conductors are exactly in phase, and the differential mode current is zero.

When $\Psi = 90°$, equation (2-9) reduces to

$$\frac{I(s,\omega)}{E_z^i(\omega)} = \frac{-j2\sin\beta b/2}{\beta D}\left[Z_0\sin\beta s + jZ_1(1 - \cos\beta s)\right]. \tag{2-10}$$

The geometry is shown in Fig. 2-9. The magnitude of equation (2-10) is plotted in Figs. 2-10 and 2-11 for two different line geometries and three load impedance combinations.

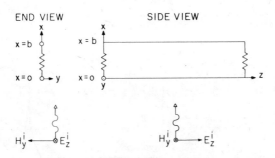

FIGURE 2-9. E_z traveling in the x direction ($\Psi = 90°$).

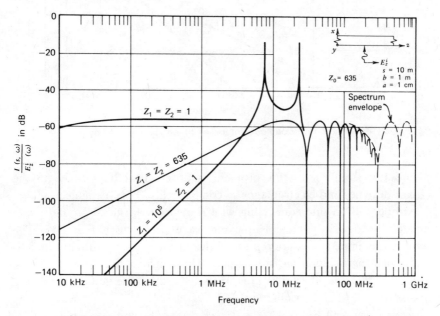

FIGURE 2-10. Load current for a 10-m-long line excited by E_z^i.

2.3 WIRE OVER A GROUND PLANE—A CIRCUIT APPLICATION

Electromagnetic fields inside electronic equipment induce noise on the interconnecting wires and cables. The fields may arise from sources inside the box (power supplies, relay transients, etc.) or from external sources (broadcast stations, induction heaters, etc.). If the induced noise exceeds the susceptibility threshold voltage of a circuit, malfunction can result. The

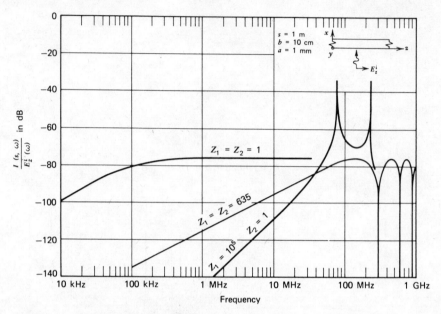

FIGURE 2-11. Load current for a 1-m-long line excited by E_z^i.

quantitative solution to this problem is of practical importance to the circuit designer and electromagnetic compatibility (EMC) engineer alike.

A model of an interconnecting wire routed over a metal ground plane and illuminated by an incident plane wave is shown in Fig. 2-12. The terminating impedances are representative of the driving and input impedances of a particular transistor circuit. The height of the wire above the ground plane is h. The problem is to find V_L, the noise voltage appearing across the right-hand termination Z_B.

By analogy with Fig. 1-4,

$$Z_1 = 2Z_A = \frac{2R_A}{1 + j\omega R_A C_A} \tag{2-11}$$

$$Z_2 = 2Z_B = \frac{2R_B}{1 + j\omega R_B C_B} \tag{2-12}$$

$$h = b/2 \text{ (height of wire above ground)}. \tag{2-13}$$

The current in the load impedance Z_B, from equation (1-12) with $E_x = 0$

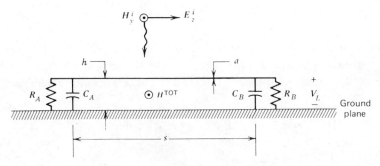

FIGURE 2-12.   Circuit wire over a metal ground plane.

and E_z uniform in the z direction (i.e., independent of z), is

$$I(s,\omega) = \frac{2K(\omega)\left[Z_0\sin\beta s + jZ_1(1-\cos\beta s)\right]}{\beta\left[(Z_0Z_1 + Z_0Z_2)\cos\beta s + j(Z_0^2 + Z_1Z_2)\sin\beta s\right]} \qquad (2\text{-}14)$$

where

$$K(\omega) = \left[2j\sin(\beta b/2)\right]E_z^i = (2j\sin\beta h)E_z^i$$
$$Z_0 = 276\log_{10}(2b/a) = 276\log_{10}(h/a). \qquad (2\text{-}15)$$

$K(\omega)$, which was derived using the method of images in Section 1.4, is simply the total electric field at a height h above the ground plane. This is shown by summing the incident and reflected fields directly, taking the phase reference at the ground plane. Thus the field at any point h above the ground plane is

$$E_z^{\text{tot}}(h) = E_z^i(h) + E_z^{\text{refl}}(h) = E_z^i\varepsilon^{+j\beta h} - E_z^i\varepsilon^{-j\beta h}.$$

or   (2-16)

$$E_z^{\text{tot}}(h) = (2j\sin\beta h)E_z^i = K(\omega).$$

As a matter of practical interest, it is generally not possible to measure the incident field. In the presence of the ground plane, the field that would be measured using a probe such as a short dipole would be the total electric field given by (2-16). But the total electric field is very small compared to the incident field when βh is small, making its measurement difficult or impossible because of practical limitations on instrument sensitivity.

On the other hand, for small βh, the total magnetic field near the surface of the ground plane is approximately twice the incident magnetic field strength. This is shown by again taking the phase reference at the ground plane. Then

$$H_y^{\text{tot}}(h) = H_y^i(h) + H_y^{\text{refl}}(h) = H_y^i \varepsilon^{+j\beta h} + H_y^i \varepsilon^{-j\beta h}$$

or   (2-17)

$$H_y^{\text{tot}}(h) = (2\cos\beta h) H_y^i.$$

From a measurement viewpoint, it is preferable to measure the total magnetic field near the surface, thus requiring that the solution be in terms of this quantity. Since the incident field is a plane wave,

$$E_z^i = 120\pi H_y^i.$$ (2-18)

Substituting this relation into (2-16) yields

$$K(\omega) = 120\pi(2j\sin\beta h) H_y^i.$$ (2-19)

Solving (2-17) for H_y^i and substituting in (2-19) results in

$$K(\omega) = \frac{120\pi(2j\sin\beta h)}{2\cos\beta h} H_y^{\text{tot}}(h).$$ (2-20)

The voltage V_L across the right-hand termination Z_b is

$$V_L(\omega) = I(s,\omega) Z_b.$$

Finally, using (2-11), (2-12), and (2-20) in equation (2-14) and performing some algebra we have

$$V_L(\omega) = H_y^{\text{tot}} \frac{j120\pi\sin\beta h}{\beta\cos\beta h} \frac{P}{Q}$$ (2-21)

where

$$P = \sin\beta s + j\left(\omega R_A C_A \sin\beta s + \frac{2R_A}{Z_0} - \frac{2R_A}{Z_0}\cos\beta s\right)$$

$$Q = \left[\left(1 + \frac{R_A}{R_B}\right)\cos\beta s - \frac{\omega Z_0}{2}\left(C_B + \frac{R_A C_A}{R_B}\right)\sin\beta s\right]$$

$$+ j\left[\omega R_A(C_A + C_B)\cos\beta s + \left(\frac{Z_0}{2R_B} - \frac{\omega^2 R_A C_A C_B Z_0}{2} + \frac{2R_A}{Z_0}\right)\sin\beta s\right].$$

For a numerical example, the following parameters are chosen:

$s = 0.5$ m   (line length)

$h = 5$ mm   (height of wire over the ground plane)

$a = 0.25$ mm   (wire diameter)

$R_A = 20$

$C_A = 10$ pf

$R_B = 100,000$

$C_B = 10$ pf.

Figure 2-13 is a plot, in decibels, of the load voltage spectrum in equation (2-21) normalized to the magnetic field strength, that is

$$20 \log_{10} \left| \frac{V_L(\omega)}{H_y^{tot}(\omega)} \right|.$$

If H_y^{tot} is expressed in dB μA/m (dB referenced to 1 microampere per meter), then V_L is in units of dB μV (dB referenced to 1 microvolt). If H_y^{tot} is expressed in dB A/m (dB referenced to 1 ampere per meter), then V_L is in units of dB V (dB referenced to 1 volt).

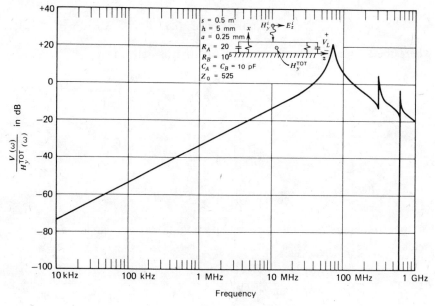

FIGURE 2-13. Load voltage for wire over ground plane in terms of total magnetic field.

2.4 TIME-DOMAIN RESPONSE

Finding the time-domain response of a transmission line excited by an electromagnetic field requires application of the inverse Fourier transform to the previously developed frequency-domain solutions, as discussed in Section 1.5. Usually, the resulting equations are too cumbersome for practical use. However, when certain simplifying conditions are satisfied, the time-domain solutions are straightforward and can provide a great deal of insight. In this section the general method for obtaining the time-domain response for the load current of a lossless line illuminated by a plane wave with arbitrary time dependence is presented in some detail. Examples are also given.

Consider a time-dependent electric field $E_x^i(t)$ polarized parallel to the terminations and traveling in the $-y$ direction as shown in Fig. 2-5. The frequency-domain solution for the current in the right-hand load Z_2 of a lossless line is given by equation (2-6) as

$$I(s,\omega) = E_x^i(\omega) \frac{b\left[Z_0(1-\cos\beta s) - jZ_1 \sin\beta s\right]}{(Z_0Z_1 + Z_0Z_2)\cos\beta s + j(Z_0^2 + Z_1Z_2)\sin\beta s}. \qquad (2\text{-}22)$$

Following Whitescarver's method (see Ref [3] for Chapter 1 in the Bibliography), this relation can be expressed as

$$I(s,\omega) = T(\omega)E_x^i(\omega) \qquad (2\text{-}23)$$

where the frequency-domain transfer function is defined by

$$T(\omega) = \frac{b\left[Z_0(1-\cos\beta s) - jZ_1 \sin\beta s\right]}{(Z_0Z_1 + Z_0Z_2)\cos\beta s + j(Z_0^2 + Z_1Z_2)\sin\beta s}. \qquad (2\text{-}24)$$

Using the identities

$$\sin x = \frac{1}{2j}(\varepsilon^{jx} - \varepsilon^{-jx})$$

$$\cos x = \frac{1}{2}(\varepsilon^{jx} + \varepsilon^{-jx})$$

for the trigonometric functions in equation (2-24) and collecting terms yields

$$T(\omega) = \frac{b\left[2Z_0 - (Z_0 + Z_1)\varepsilon^{j\beta s} - (Z_0 - Z_1)\varepsilon^{-j\beta s}\right]}{(Z_0Z_1 + Z_0Z_2 + Z_0^2 + Z_1Z_2)\varepsilon^{j\beta s} - (Z_0^2 + Z_1Z_2 - Z_0Z_1 - Z_0Z_2)\varepsilon^{-j\beta s}}.$$

Multiplying numerator and denominator by $\varepsilon^{-j\beta s}$, factoring out the term $(Z_0 + Z_1)$ from numerator and denominator, and factoring out $(Z_0 + Z_2)$ from the denominator leads to

$$T(\omega) = \frac{-b}{2Z_0} \frac{2Z_0}{(Z_0 + Z_2)} \frac{\left[1 - \dfrac{2Z_0}{Z_0 + Z_1}\varepsilon^{-j\beta s} + \dfrac{Z_0 - Z_1}{Z_0 + Z_1}\varepsilon^{-j2\beta s}\right]}{1 - \dfrac{(Z_0 - Z_1)(Z_0 - Z_2)}{(Z_0 + Z_1)(Z_0 + Z_2)}\varepsilon^{-j2\beta s}}. \qquad (2\text{-}25)$$

From transmission line theory, the sending- and receiving-end reflection coefficients for current and voltage are given by

$$\rho_1^I = \frac{Z_0 - Z_1}{Z_0 + Z_1} \qquad \text{sending-end current reflection coefficient}$$

$$\rho_2^I = \frac{Z_0 - Z_2}{Z_0 + Z_2} \qquad \text{receiving-end current reflection coefficient}$$

$$\qquad (2\text{-}26)$$

$$\rho_1^v = -\rho_1^I = \frac{Z_1 - Z_0}{Z_0 + Z_1} \qquad \text{sending-end voltage reflection coefficient}$$

$$\rho_2^v = -\rho_2^I = \frac{Z_2 - Z_0}{Z_0 + Z_2} \qquad \text{receiving-end voltage reflection coefficient.}$$

Similarly, transmission coefficients are defined as

$$T_1^I = 1 - \rho_1^I = \frac{2Z_1}{Z_0 + Z_1}$$

$$T_2^I = 1 - \rho_2^I = \frac{2Z_2}{Z_0 + Z_2}$$

$$\qquad (2\text{-}27)$$

$$T_1^v = 1 - \rho_1^v = \frac{2Z_0}{Z_0 + Z_1}$$

$$T_2^v = 1 - \rho_2^v = \frac{2Z_0}{Z_0 + Z_2}.$$

Equation (2-25) can be rewritten as

$$T(\omega) = \frac{-b}{2Z_0} T_2^v \frac{\left(1 - T_1^v \varepsilon^{-j\beta s} + \rho_1^I \varepsilon^{-j2\beta s}\right)}{1 - \rho_1^I \rho_2^I \varepsilon^{-j2\beta s}}. \tag{2-28}$$

When the terminating impedances are real, the reflection and transmission coefficients are real and independent of frequency. The denominator of (2-28) can then be expanded using the binomial theorem:

$$(1-x)^{-1} = 1 + x + x^2 + \cdots = \sum_{n=0}^{\infty} x^n \qquad |x| < 1.$$

The result is

$$T(\omega) = \frac{-b}{2Z_0} T_2^v \left(1 - T_1^v \varepsilon^{-j\beta s} + \rho_1^I \varepsilon^{-j2\beta s}\right) \sum_{n=0}^{\infty} \left(\rho_1^I \rho_2^I\right)^n \varepsilon^{-j2n\beta s}.$$

The time-domain load current from equation (2-23) is (see Section 1.5)

$$I(s,t) = \frac{1}{2\pi} \int_{-\infty}^{\infty} E_x^i(\omega) T(\omega) \varepsilon^{j\omega t} \, d\omega$$

or

$$I(s,t) = \frac{1}{2\pi} \int_{-\infty}^{\infty} E_x^i(\omega) \left[\frac{-b}{2Z_0} \left(1 - T_1^v \varepsilon^{-j\beta s} + \rho_1^I \varepsilon^{-j2\beta s}\right) \right]$$

$$\times \sum_{n=0}^{\infty} \left(\rho_1^I \rho_2^I\right)^n \varepsilon^{-j2n\beta s} \varepsilon^{j\omega t} \, d\omega.$$

Making the substitution $\beta s = \omega s / c$ and factoring out the terms that are independent of ω results in

$$I(s,t) = \frac{-b}{2Z_0} T_2^v \sum_{n=0}^{\infty} \left(\rho_1^I \rho_2^I\right)^n \left\{ \frac{1}{2\pi} \int_{-\infty}^{\infty} E_x^i(\omega) \right.$$

$$\left. \times \left[\varepsilon^{-j2n\omega s/c} - T_1^v \varepsilon^{-j\omega(s/c + 2ns/c)} + \rho_1^I \varepsilon^{-j\omega(2s/c + 2ns/c)}\right] \varepsilon^{j\omega t} \, d\omega \right\}.$$

Using the shifting theorem, we have

$$I(s,t) = \frac{-b}{2Z_0} T_2^v \sum_{n=0}^{\infty} \left(\rho_1^I \rho_2^I\right)^n$$

$$\times \left[E_x^i(t - 2ns/c) - T_1^v E_x^i(t - s/c - 2ns/c) + \rho_1^I E_x^i(t - 2s/c - 2ns/c) \right]$$

$$\tag{2-29}$$

which is the desired expression for the time-domain current in the load. Note that (2-29) is, in general, the sum of an infinite number of terms containing the time-shifted incident electric field modified by the reflection and transmission coefficients at the ends of the line. This is the expected result for an unmatched line. That is, the response excited on the line continually reflects back and forth between the terminations, undergoing modification at each reflection. When either end of the line is matched, (2-29) has only a finite number of terms, and the time-domain solution is relatively easy to obtain. Two examples will illustrate this.

For the first example, consider a 90-m-long line matched at both ends. Then

$$s = 90 \text{ m}$$

$$Z_1 = Z_2 = Z_0.$$

From equations (2-26) and (2-27),

$$\rho_1^I = \rho_2^I = 0$$

$$T_1^v = T_2^v = 1.$$

Also, $s/c = 90/3 \times 10^8 = 300$ ns (nanoseconds).

Under the matched conditions, (2-29) is nonzero only for $n = 0$. The result is

$$I(s,t) = \frac{b}{2Z_0} \left[-E_x^i(t) + E_x^i(t - 300 \text{ ns}) \right] \qquad (2\text{-}30)$$

which is shown in Fig. 2-14 for some arbitrary time dependent electric field.

The mathematical result for this example can be readily verified by referring again to Fig. 2-5. The electric field arrives at both terminations simultaneously and induces a voltage $bE(t)$. A current i_2 immediately flows in the right-hand termination. From Fig. 2-15, this current is seen to be equal to $bE(t)/2Z_0$ and flows in a direction opposite to the assigned direction for $I(s,t)$. This is the first term of equation (2-30). The current i_2 then propagates toward the left-hand termination, where it is completely absorbed, because the line is matched at that end ($Z_1 = Z_0$). The current i_1 induced in the left-hand termination after $t = 0$ propagates toward Z_2 and arrives after a delay of $s/c = 300$ nanoseconds. This is the second term of equation (2-30).

The same 90-m-long line with the left-hand termination short-circuited

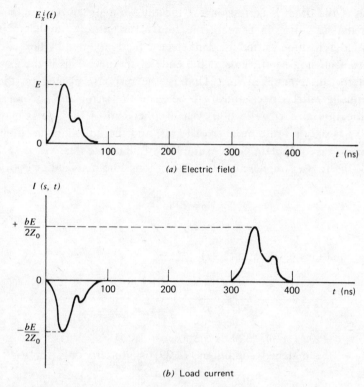

(a) Electric field

(b) Load current

FIGURE 2-14. Time-domain solution for matched line example.

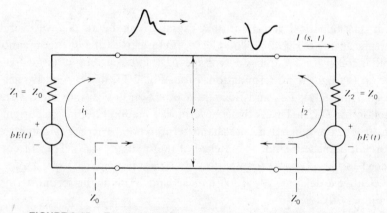

FIGURE 2-15. Transmission line analysis of matched line example.

will serve as a second example. Now

$$Z_1 = 0$$

$$Z_2 = Z_0$$

and, from (2-26) and (2-27)

$$\rho_1^I = 1$$

$$\rho_2^I = 0$$

$$T_1^v = 2$$

$$T_2^v = 1.$$

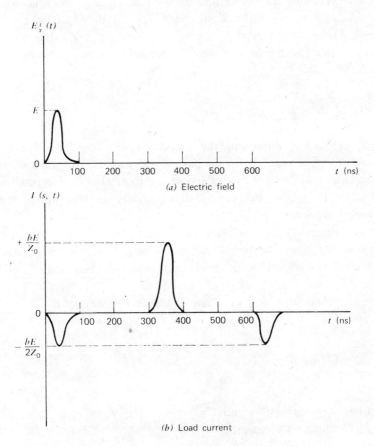

(a) Electric field

(b) Load current

FIGURE 2-16. Time-domain solution for line shorted at one end.

Equation (2-29) is again nonzero only for $n = 0$, and we have

$$I(s,t) = \frac{b}{2Z_0} \left[-E_x^i(t) + 2E_x^i(t - 300 \text{ ns}) - E_x^i(t - 600 \text{ ns}) \right]$$

which is plotted in Fig. 2-16 for the arbitrary incident field shown. Again, these waveforms are readily verified. The voltage $bE(t)$ induced in the right-hand termination after $t = 0$ immediately causes a current $-bE(t)/2Z_0$ in Z_2. This current travels to the left-hand termination, where it is completely reflected at the short and returns to the right-hand load after a delay of 600 ns (since the wave had to travel twice the length of the line). The voltage $bE(t)$ induced in the left-hand termination after $t = 0$ injects a current on the line equal to $bE(t)/Z_0$, which is twice the magnitude of the current injected at the right-hand end of the line (since $Z_1 = 0$). This current travels to the right and arrives at Z_2 after a delay of 300 ns. Since the line is matched at this end, no more reflections take place.

2.5 DISSIPATIVE LINE RESPONSE

In this section, the solution for the load current of a lossy line is developed. This is undertaken mainly for the purpose of verifying that the lossless line assumption used in most radio frequency transmission line problems is indeed a practical one, in the sense that it greatly simplifies the mathematics and at the same time leads to acceptably accurate answers.

The illuminating field selected is a plane wave E_z traveling in the x direction as shown in Fig. 2-9. For a lossy line, the current in the right-hand load impedance Z_2 is given by equation (1-3), with $E_x^i = 0$, as

$$I(s,\omega) = \frac{1}{D} \int_0^s K(z,\omega) \left[Z_0 \cosh \gamma z + Z_1 \sinh \gamma z \right] dz \qquad (2\text{-}31)$$

where

$K(z,\omega) = E_z^i(b,z,\omega) - E_z^i(0,z,\omega)$
$D = (Z_0 Z_1 + Z_0 Z_2) \cosh \gamma s + (Z_0^2 + Z_1 Z_2) \sinh \gamma s$
$\gamma = \alpha + j\beta$
α = attenuation constant of line
$\beta = 2\pi/\lambda$ phase constant of line
$Z_0 = \sqrt{Z/Y}$ characteristic impedance.

Since the incident plane wave field is uniform over each conductor, $K(z,\omega)$ is independent of z and is given by equation (2-8) with $\psi = 90°$, that is

$$K(\omega) = E_z^i(\omega)\left[-j2\sin\frac{\beta b}{2}\right]. \tag{2-32}$$

Then (2-31) can be rewritten as

$$I(s,\omega) = \frac{E_z^i(\omega)\left[-j2\sin\dfrac{\beta b}{2}\right]}{D}\int_0^s\left[Z_0\cosh\gamma z + Z_1\sinh\gamma z\right]dz. \tag{2-33}$$

Performing the integration results in

$$I(s,\omega) = \frac{E_z^i(\omega)\left[-j2\sin\dfrac{\beta b}{2}\right]\left[Z_0\sinh\gamma s + Z_1(\cosh\gamma s - 1)\right]}{\gamma\left[(Z_0Z_1 + Z_0Z_2)\cosh\gamma s + (Z_0^2 + Z_1Z_2)\sinh\gamma s\right]}. \tag{2-34}$$

In order to arrive at numerical results, the hyperbolic functions with complex arguments in (2-34) must be expressed in terms of functions with real arguments. Using the identities

$$\cosh\gamma z = \cosh\alpha z\cos\beta z + j\sinh\alpha z\sin\beta z$$

$$\sinh\gamma z = \sinh\alpha z\cos\beta z + j\cosh\alpha z\sin\beta z$$

we have

$$I(s,\omega) = \frac{E_z^i(\omega)\left[-j2\sin\dfrac{\beta b}{2}\right]}{\gamma}\times\frac{\text{NUM}}{\text{DEN}} \tag{2-35}$$

where
$$\begin{aligned}\text{NUM} =\ &(Z_0\sinh\alpha s\cos\beta s + Z_1\cosh\alpha s\cos\beta s - Z_1)\\&+ j(Z_0\cosh\alpha s\sin\beta s + Z_1\sinh\alpha s\sin\beta s)\end{aligned}$$
and
$$\begin{aligned}\text{DEN} =\ &\left[(Z_0Z_1 + Z_0Z_2)(\cosh\alpha s\cos\beta s)\right.\\&\left. + (Z_0^2 + Z_1Z_2)(\sinh\alpha s\cos\beta s)\right]\\& + j\left[(Z_0Z_1 + Z_0Z_2)(\sinh\alpha s\sin\beta s)\right.\\&\left. + (Z_0^2 + Z_1Z_2)\cosh\alpha s\sin\beta s\right].\end{aligned}$$

For a lossy line, the characteristic impedance Z_0 is, in general, complex.

That is,

$$Z_0 = \sqrt{\frac{Z}{Y}} = \sqrt{\frac{R + j\omega L}{G + j\omega C}}$$

where R, G, L, and C are, respectively, the resistance, conductance, inductance, and capacitance per unit length of the line. In the voice frequency band (300 to 3000 Hz) and below, the complex characteristic impedance must be used in equation (2-35). However, at higher frequencies we have $\omega L \gg R$ and $\omega C \gg G$, so that the characteristic impedance is, to a good approximation, real valued and is given by

$$Z_0 = \sqrt{\frac{L}{C}} = 120 \ln \frac{2b}{a} = 276 \log_{10} \frac{2b}{a} \qquad (2\text{-}36)$$

where b is the conductor spacing and a is the conductor diameter. Since Z_0 can be taken as real, computations using (2-35) are considerably simplified.

In order to arrive at numerical results with minimum mathematical manipulations, assume the line is matched at both ends, that is, $Z_1 = Z_2 = Z_0$. Then equation (2-35) reduces to

$$I(s,\omega) = \frac{E_z^i(\omega)\left[-j2\sin\beta b/2\right]}{2\gamma Z_0}$$

$$\times \frac{\left[(\sinh\alpha s + \cosh\alpha s)\cos\beta s - 1\right] + j(\sinh\alpha s + \cosh\alpha s)\sin\beta s}{(\sinh\alpha s + \cosh\alpha s)\cos\beta s + j(\sinh\alpha s + \cosh\alpha s)\sin\beta s}$$

or

$$I(s,w) = \frac{E_z^i(\omega)\left[-j2\sin\beta b/2\right]\varepsilon^{-j\beta s}}{2\gamma Z_0}$$

$$\times \frac{\left[(\sinh\alpha s + \cosh\alpha s)\cos\beta s - 1\right] + j(\sinh\alpha s + \cosh\alpha s)\sin\beta s}{\sinh\alpha s + \cosh\alpha s}.$$

Define a transfer function T as the magnitude of the load current normalized to the incident field. That is,

$$T = \left| \frac{I(s,\omega)}{E_z^i(\omega)} \right|.$$

Then

$$T = \frac{\left|2\sin\dfrac{\beta b}{2}\right|}{2Z_0(\alpha^2+\beta^2)^{1/2}}$$

$$\times \frac{\left\{\left[(\sinh\alpha s + \cosh\alpha s)\cos\beta s - 1\right]^2 + (\sinh\alpha s + \cosh\alpha s)^2 \sin^2\beta s\right\}^{1/2}}{\sinh\alpha s + \cosh\alpha s}$$

or

$$T = \frac{\left|2\sin\dfrac{\beta b}{2}\right|}{2Z_0(\alpha^2+\beta^2)^{1/2}}$$

$$\times \frac{\left\{(\sinh\alpha s + \cosh\alpha s)^2 - 2(\sinh\alpha s + \cosh\alpha s)\cos\beta s + 1\right\}^{1/2}}{\sinh\alpha s + \cosh\alpha s}. \qquad (2\text{-}37)$$

Figure 2-17 shows the load current spectrum of a 10-m-long lossy line

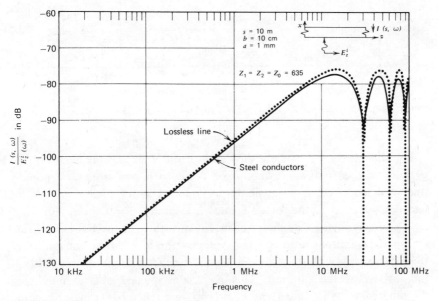

FIGURE 2-17. Load current for a 10-m-long steel conductor line illuminated by E_z^i.

with steel conductors. The load current was calculated from equation (2-37) using the values of the attenuation constant α plotted in Fig. 1-6. In Fig. 2-18, the load current for a 100-m-long steel conductor line is plotted. In both figures, the conductor spacing is 10 cm and the conductor diameter is 1 mm. The lossless line current spectrum, calculated from equation (2-10) in Section 2.2, is also shown in Figs. 2-17 and 2-18 for comparison.

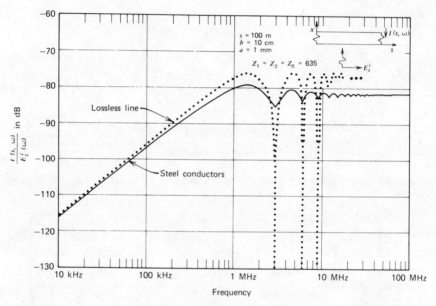

FIGURE 2-18.   Load current for a 100-m-long steel conductor line illuminated by E_z^i.

The dissipative losses cause the depth of the nulls and the height of the peaks in the current spectrums to be diminished compared to a lossless line, and the effect is increasingly pronounced at the higher frequencies (since α increases as the square root of frequency).

It is interesting to note that the current spectrum for the 100-m-long line in Fig. 2-18 approaches a constant value at the higher frequencies, rather than decreasing with frequency, as would be the case with a transmission line driven at the far end. This is a consequence of the line being illuminated along its entire length and can be verified by letting αs become large in equation (2-37). Factoring out $\cosh \alpha s$ from the numerator and denominator of (2-37) yields

$$T = \frac{\left|2\sin\frac{\beta b}{2}\right|}{2Z_0(\alpha^2+\beta^2)^{1/2}}$$

$$\times \frac{\left\{(1+\tanh\alpha s)^2 - 2\frac{(1+\tanh\alpha s)}{\cosh\alpha s} + \frac{1}{\cosh^2\alpha s}\right\}^{1/2}}{1+\tanh\alpha s}.$$

But $\tanh\infty = 1$ and $\cosh\infty = \infty$. Then

$$T = \frac{\left|\sin\frac{\beta b}{2}\right|}{Z_0(\alpha^2+\beta^2)^{1/2}}. \tag{2-38}$$

At the higher frequencies

$$\beta = \frac{2\pi}{\lambda} \gg \alpha.$$

Also, for conductor spacings much less than a wavelength

$$\sin\frac{\beta b}{2} \doteq \frac{\beta b}{2}.$$

Then (2-38) becomes

$$T = \frac{b}{2Z_0}.$$

For the examples in Figs. 2-17 and 2-18

$$b = 0.1 \text{ m}$$

and

$$Z_0 = 635 \text{ ohms.}$$

We have

$$T = \frac{0.1}{2\times635} = \frac{1}{12,700}$$

or

$$T(\text{dB}) = 20\log_{10}\frac{1}{12,700} = -82 \text{ dB}$$

which is exactly the high-frequency limit shown in Fig. 2-18.

CHAPTER
THREE

EXCITATION BY NONUNIFORM FIELDS

When the source of an electromag-
netic field is in close proximity to a transmission line, the fields that
illuminate the line are highly nonuniform, and the plane wave solutions of
the previous chapter are not applicable. In this chapter, the general
equations of Chapter 1 are solved for two specific nonuniform fields: those
produced by a small loop and those produced by a short dipole.

The exact formulas even for these elementary sources contain complicated integrals that require the use of a computer to produce numerical
results. For this reason, an approximate solution which greatly reduces the
computational effort is derived for the small loop in Section 3.2. The
approximate method produces results which agree quite closely with the
exact solution, and is applicable to other nonuniform illuminating fields as
well.

3.1 EXACT EQUATIONS FOR A LINE EXCITED BY A SMALL LOOP†

Figure 3-1 shows the two-wire transmission line and a small, coplanar loop
antenna that excites the line. The coordinates for the transmission line are
(x,y,z), and the coordinates for the loop are (r,θ). The conductors of the
line are spaced a distance b apart and are of diameter a. The length of the
line is s. Z_1 and Z_2 are the terminating impedances, and Z_0 is the
characteristic impedance of the line. The perpendicular distance from the
loop to a point midway between the conductors $(x = b/2)$ is R, and the
loop is adjacent to the line at a distance w from the beginning of the line.
$E_\theta(r,\omega)$ is the spatio-spectral electric field in the x–z plane that illuminates
the line, and $I(s,\omega)$ is the resulting current spectrum in the load Z_2 at $z = s$.

The loop produces a transverse electric field with a $1/r^2$ induction field
attenuation rate and a $1/r$ radiation field attenuation rate, where r is the

†Sections 3.1 to 3.3 are based on material from "The Response of a Two-Wire Transmission
Line Excited by the Nonuniform Electromagnetic Fields of a Nearby Loop" by A. A. Smith,
Jr., *IEEE Transactions on Electromagnetic Compatibility*, Vol. EMC-16, No. 4, November,
1974. Copyright 1974 by the Institute of Electrical and Electronics Engineers, Inc.

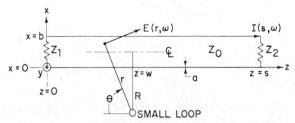

FIGURE 3-1. Geometry for transmission line and loop.

radial distance from the loop. The magnitude of the transverse field is independent of the azimuth angle θ from the loop.

The solution proceeds from equation (1-6) for dissipationless lines:

$$I(s,\omega) = \frac{1}{D} \int_0^s K(z,\omega)\left[Z_0 \cos \beta z + jZ_1 \sin \beta z\right] dz$$

$$+ \frac{Z_0}{D} \int_0^b E_x^i(x,0,\omega)\,dx - \frac{Z_0 \cos \beta s + jZ_1 \sin \beta s}{D} \int_0^b E_x^i(x,s,\omega)\,dx$$

$$(3\text{-}1)$$

where

$D = (Z_0 Z_1 + Z_0 Z_2)\cos \beta s + j(Z_0^2 + Z_1 Z_2)\sin \beta s$
$K(z,\omega) = E_z^i(b,z,\omega) - E_z^i(0,z,\omega)$
$E_z^i(b,z,\omega) = $ field in z direction incident on upper conductor $(x=b)$
$E_z^i(0,z,\omega) = $ field in z direction incident on lower conductor $(x=0)$
$E_x^i(x,0,\omega) = $ field in x direction incident on left-hand termination $(z=0)$
$E_x^i(x,s,\omega) = $ field in x direction incident on right-hand termination $(z=s)$
$\beta = 2\pi/\lambda$ phase constant
$\lambda = $ wavelength
$Z_0 = 276 \log_{10}(2b/a)$ characteristic impedance
$b = $ spacing (meters)
$a = $ diameter (meters).

For convenience, (3-1) is normalized to $E(R,\omega)$, which is the field incident on the line at $z=w$, $x=b/2$. This is done to make the solution dependent on the incident field, which can be measured or calculated, rather than on the parameters of the loop antenna. Then

$$\frac{I(s,\omega)}{E(R,\omega)} = \frac{1}{D} \int_0^s \frac{E_z^i(b,z,\omega)}{E(R,\omega)} \left[Z_0 \cos \beta z + jZ_1 \sin \beta z \right] dz$$

$$- \frac{1}{D} \int_0^s \frac{E_z^i(0,z,\omega)}{E(R,\omega)} \left[Z_0 \cos \beta z + jZ_1 \sin \beta z \right] dz$$

$$+ \frac{Z_0}{D} \int_0^b \frac{E_x^i(x,0,\omega)}{E(R,\omega)} dx$$

$$- \frac{Z_0 \cos \beta s + jZ_1 \sin \beta s}{D} \int_0^b \frac{E_x^i(x,s,\omega)}{E(R,\omega)} dx \qquad (3\text{-}2)$$

where now

$$\frac{K(z,\omega)}{E(R,\omega)} = \frac{E_z^i(b,z,\omega)}{E(R,\omega)} - \frac{E_z^i(0,z,\omega)}{E(R,\omega)} .$$

The field in the plane of a small loop is given by Schelkunoff and Friis as (see reference [1] in the Bibliography)

$$E(r,\omega) = \frac{jM(\omega)\mu\omega^3}{c^2} \left[\frac{j}{\beta r} + \frac{1}{(\beta r)^2} \right] \varepsilon^{-j\beta r} \qquad (3\text{-}3)$$

where $M(\omega)$ is a function of the loop current spectrum and loop area. If (3-3) is normalized to $E(R,\omega)$ and separated into the induction field ($\beta r \ll 1$) and radiation field ($\beta r \gg 1$) components,

$$\frac{E(r,\omega)}{E(R,\omega)} = \begin{cases} \left(\dfrac{R}{r}\right)^2 \varepsilon^{-j\beta(r-R)} & \text{Induction} \\[2ex] \dfrac{R}{r} \varepsilon^{-j\beta(r-R)} & \text{Radiation.} \end{cases} \qquad (3\text{-}4)$$

Separating the field into induction and radiation components simplifies the mathematics and leaves only a small region of uncertainty in the solution around $\beta r = 1$.

From Fig. 3-1,

$$E_z^i(x,z,\omega) = E(r,\omega)\sin\theta$$

and   (3-5)

$$E_x^i(x,z,\omega) = E(r,\omega)\cos\theta.$$

Then, normalizing (3-5) to $E(R,\omega)$ and using (3-4),

$$\frac{E_z^i(x,z,\omega)}{E(R,\omega)} = \begin{cases} \left(\dfrac{R}{r}\right)^2 \sin\theta\, \varepsilon^{-j\beta(r-R)} & \text{Induction} \\[3mm] \dfrac{R}{r} \sin\theta\, \varepsilon^{-j\beta(r-R)} & \text{Radiation} \end{cases} \tag{3-6}$$

and

$$\frac{E_x^i(x,z,\omega)}{E(R,\omega)} = \begin{cases} \left(\dfrac{R}{r}\right)^2 \cos\theta\, \varepsilon^{-j\beta(r-R)} & \text{Induction} \\[3mm] \dfrac{R}{r} \cos\theta\, \varepsilon^{-j\beta(r-R)} & \text{Radiation.} \end{cases} \tag{3-7}$$

On the upper conductor of the transmission line (at $x = b$)

$$h_1(z) = \frac{R}{r} = \frac{R}{\sqrt{(w-z)^2+(R+b/2)^2}}$$

$$h_2(z) = \sin\theta = \frac{R+b/2}{\sqrt{(w-z)^2+(R+b/2)^2}}$$

$$h_3(z) = r - R = \sqrt{(w-z)^2+(R+b/2)^2} - R.$$

On the lower conductor of the transmission line (at $x = 0$)

$$h_4(z) = \frac{R}{r} = \frac{R}{\sqrt{(w-z)^2+(R-b/2)^2}}$$

$$h_5(z) = \sin\theta = \frac{R-b/2}{\sqrt{(w-z)^2+(R-b/2)^2}}$$

$$h_6(z) = r - R = \sqrt{(w-z)^2+(R-b/2)^2} - R.$$

On the left-hand termination (at $z=0$)

$$h_7(x) = \frac{R}{r} = \frac{R}{\sqrt{(R-b/2+x)^2+w^2}}$$

$$h_8(x) = \cos\theta = \frac{w}{\sqrt{(R-b/2+x)^2+w^2}}$$

$$h_9(x) = r - R = \sqrt{(R-b/2+x)^2+w^2} - R.$$

On the right-hand termination (at $z=s$)

$$h_{10}(x) = \frac{R}{r} = \frac{R}{\sqrt{(R-b/2+x)^2+(s-w)^2}}$$

$$h_{11}(x) = \cos\theta = \frac{-(s-w)}{\sqrt{(R-b/2+x)^2+(s-w)^2}}$$

$$h_{12}(x) = r - R = \sqrt{(R-b/2+x)^2+(s-w)^2} - R.$$

Substituting the $h(z)$ and $h(x)$ functions defined above into (3-6) and (3-7), we have

$$\frac{E_z^i(b,z,\omega)}{E(R,\omega)} = \begin{cases} h_1^2 h_2(\cos\beta h_3 - j\sin\beta h_3) & \text{Induction} \\ h_1 h_2(\cos\beta h_3 - j\sin\beta h_3) & \text{Radiation} \end{cases} \qquad (3\text{-}8)$$

$$\frac{E_z^i(0,z,\omega)}{E(R,\omega)} = \begin{cases} h_4^2 h_5(\cos\beta h_6 - j\sin\beta h_6) & \text{Induction} \\ h_4 h_5(\cos\beta h_6 - j\sin\beta h_6) & \text{Radiation} \end{cases} \qquad (3\text{-}9)$$

$$\frac{E_x^i(x,0,\omega)}{E(R,\omega)} = \begin{cases} h_7^2 h_8(\cos\beta h_9 - j\sin\beta h_9) & \text{Induction} \\ h_7 h_8(\cos\beta h_9 - j\sin\beta h_9) & \text{Radiation} \end{cases} \qquad (3\text{-}10)$$

$$\frac{E_x^i(x,s,\omega)}{E(R,\omega)} = \begin{cases} h_{10}^2 h_{11}(\cos\beta h_{12} - j\sin\beta h_{12}) & \text{Induction} \\ h_{10} h_{11}(\cos\beta h_{12} - j\sin\beta h_{12}) & \text{Radiation.} \end{cases} \qquad (3\text{-}11)$$

Substituting the induction field terms from (3-8) through (3-11) into

(3-2), yields

$$\frac{I(s,\omega)}{E(R,\omega)} = \frac{1}{D}\left[Z_0 \int_0^s h_1^2 h_2 \cos \beta z \cos \beta h_3 \, dz \right.$$

$$-jZ_0 \int_0^s h_1^2 h_2 \cos \beta z \sin \beta h_3 \, dz$$

$$+jZ_1 \int_0^s h_1^2 h_2 \sin \beta z \cos \beta h_3 \, dz$$

$$+ Z_1 \int_0^s h_1^2 h_2 \sin \beta z \sin \beta h_3 \, dz$$

$$- Z_0 \int_0^s h_4^2 h_5 \cos \beta z \cos \beta h_6 \, dz$$

$$+jZ_0 \int_0^s h_4^2 h_5 \cos \beta z \sin \beta h_6 \, dz$$

$$-jZ_1 \int_0^s h_4^2 h_5 \sin \beta z \cos \beta h_6 \, dz$$

$$- Z_1 \int_0^s h_4^2 h_5 \sin \beta z \sin \beta h_6 \, dz$$

$$+ Z_0 \int_0^b h_7^2 h_8 \cos \beta h_9 \, dx - jZ_0 \int_0^b h_7^2 h_8 \sin \beta h_9 \, dx$$

$$- Z_0 \cos \beta s \int_0^b h_{10}^2 h_{11} \cos \beta h_{12} \, dx$$

$$+ jZ_0 \cos \beta s \int_0^b h_{10}^2 h_{11} \sin \beta h_{12} \, dx$$

$$- jZ_1 \sin \beta s \int_0^b h_{10}^2 h_{11} \cos \beta h_{12} \, dx$$

$$\left. - Z_1 \sin \beta s \int_0^b h_{10}^2 h_{11} \sin \beta h_{12} \, dx \right] \tag{3-12}$$

or

$$\frac{I(s,\omega)}{E(R,\omega)} = \frac{1}{D}(G_1 - jG_2 + jG_3 + G_4 - G_5 + jG_6 - jG_7 - G_8 + G_9$$

$$- jG_{10} - G_{11} + jG_{12} - jG_{13} - G_{14}) \qquad (3\text{-}13)$$

where each G is defined by the corresponding term in (3-12). For Z_1 and Z_2 real, the magnitude of (3-13), the induction field solution, is

$$\left|\frac{I(s,\omega)}{E(R,\omega)}\right| = \frac{1}{|D|}\left[(G_1 + G_4 - G_5 - G_8 + G_9 - G_{11} - G_{12})^2\right.$$

$$\left. + (G_3 - G_2 + G_6 - G_7 - G_{10} + G_{12} - G_{13})^2\right]^{1/2}.$$

$$(3\text{-}14)$$

For the radiation field

$$\frac{I(s,\omega)}{E(R,\omega)} = \frac{1}{D}(H_1 - jH_2 + jH_3 + H_4 - H_5 + jH_6 - jH_7 - H_8 + H_9$$

$$- jH_{10} - H_{11} + jH_{12} - jH_{13} - H_{14})$$

where each H is defined by the corresponding term in (3-12) using h_1, h_4, h_7, and h_{10} instead of h_1^2, h_4^2, h_7^2, and h_{10}^2 [see (3-8) through (3-11)]. The magnitude of the radiation field solution (Z_1 and Z_2 real) is

$$\left|\frac{I(s,\omega)}{E(R,\omega)}\right| = \frac{1}{|D|}\left[(H_1 + H_4 - H_5 - H_8 + H_9 - H_{11} - H_{14})^2\right.$$

$$\left. + (H_3 - H_2 + H_6 - H_7 - H_{10} + H_{12} - H_{13})^2\right]^{1/2}. \quad (3\text{-}15)$$

If a coupling transfer function T is defined as

$$T = 20\log_{10}\left|\frac{I(s,\omega)}{E(R,\omega)}\right| \qquad (3\text{-}16)$$

then

$$I(s,\omega) \ (\text{dB } \mu\text{A}/\text{MHz}) = E(R,\omega) \ (\text{dB } \mu\text{V}/\text{m}/\text{MHz}) + T \quad (3\text{-}17)$$

where T is found from (3-14) for the induction field and from (3-15) for the radiation field.

Computer evaluation of the integrals in (3-12) requires significant time. For example, when programmed in APL/360 and run on an IBM System/360 Model 85, typical computations require approximately 5 sec for each frequency. The approximate solution developed in the next section significantly reduces the computation time.

3.2 APPROXIMATE METHOD FOR A LINE EXCITED BY A SMALL LOOP

It is the complexity of the function $K(z)$ in (3-1), due to the nonuniform amplitude and phase of the fields illuminating the line, that results in the complicated integrals indicated in (3-12). However, when the loop is near the line, the function $K(z)$ is concentrated around $z = w$ since the fields in proximity to the loop attenuate rapidly with distance. Consider, then, a uniform approximation for $K(z)$ that has a constant value of $K(w)$ over a segment of the line of length L and is zero elsewhere. (From the definition of $K(z)$ in (3-1), it is evident that this amounts to approximating the incident field by a field that is nonuniform in the x direction and is uniform or constant in the z direction over a segment of line of length L, and zero for all other z.) This is illustrated in Fig. 3-2. Then the first term of (3-1), normalized to $E(R,\omega)$, becomes

$$\frac{I_1(s,\omega)}{E(R,\omega)} = \frac{1}{D} \frac{K(w,\omega)}{E(R,\omega)} \int_{w-L/2}^{w+L/2} \left[Z_0 \cos \beta z + jZ_1 \sin \beta z \right] dz$$

$$= \frac{K(w,\omega)}{E(R,\omega)} \frac{2 \sin \beta L/2}{D\beta} \left[Z_0 \cos \beta w + jZ_1 \sin \beta w \right]. \quad (3\text{-}18)$$

Using (3-8) and (3-9) with $z = w$, and after some algebraic manipulation,

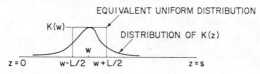

FIGURE 3-2. Distribution of $K(z)$ and an equivalent uniform distribution.

we have

$$
\frac{K(w,\omega)}{E(R,\omega)} = \begin{cases} \left(-2j\sin\frac{\beta b}{2}\right)\dfrac{1+\left(\dfrac{b}{2R}\right)^2 - j\dfrac{b}{R}\cot\dfrac{\beta b}{2}}{\left[1-\left(\dfrac{b}{2R}\right)^2\right]^2} & \text{Induction} \\[4ex] \left(-2j\sin\frac{\beta b}{2}\right)\dfrac{1-j\dfrac{b}{2R}\cot\dfrac{\beta b}{2}}{1-\left(\dfrac{b}{2R}\right)^2} & \text{Radiation.} \end{cases}
$$

Equation (3-18) can then be rewritten as

$$
\frac{I_1(s,\omega)}{E(R,\omega)} = \left(-2j\sin\frac{\beta b}{2}\right)\frac{2\sin\beta L/2}{D\beta}\left[Z_0\cos\beta w + jZ_1\sin\beta w\right]P \quad (3\text{-}19)
$$

where

$$
P = \begin{cases} \dfrac{1+\left(\dfrac{b}{2R}\right) - j\dfrac{b}{R}\cot\dfrac{\beta b}{2}}{\left[1-\left(\dfrac{b}{2R}\right)^2\right]^2} & \text{Induction} \\[4ex] \dfrac{1-j\dfrac{b}{2R}\cot\dfrac{\beta b}{2}}{1-\left(\dfrac{b}{2R}\right)^2} & \text{Radiation.} \end{cases} \quad (3\text{-}20)
$$

Equation (3-19), with $P=1$, is identical to the solution for a plane wave that illuminates a segment of line of length L centered at $z=w$. Then P, defined by (3-20), can be thought of as a factor that corrects the plane-wave solution to account for the amplitude differential of the field across the two conductors of the transmission line. (Although a plane wave induces a transmission-line mode current on the line by virtue of the phase difference of the wave on the two conductors, a nearby source induces a greater current by virtue of both a phase difference and an amplitude difference due to the finite attenuation of the field.)

The foregoing approximation assumes that the major part of the field distribution $K(z)$ illuminates the conductors as shown in Fig. 3-2 (i.e., only the minor tails of the distribution overlap the ends of the line). Under this condition, it is found that the length L of the equivalent uniform distribution is equal to the distance R from the loop to the point midway between the conductors. That is, the current in the load due to E_z is found by using $L = R$ in (3-19). The condition on the use of (3-19) that the major part of the field distribution lies on the line is then seen to be satisfied when both $R < s$ and $w > R/2$. [If the loop is located near either end of the line so that $w \approx 0$ or $w \approx s$, only one-half the distribution of $K(z)$ shown in Fig. 3-2 illuminates the line. In this case the current in the load due to E_z can be found from (3-19) with $L = R/2$, and with $w = R/4$ for the loop near the left-hand termination or with $w = s - R/4$ for the loop near the right-hand termination.]

When the loop is located far enough from the transmission line ($R \gg s$) so that E_z is approximately uniform over the entire line in the z direction (but nonuniform in the x direction due to the attenuation of the fields), use $L = s$ and $w = s/2$ in (3-19). In the limit, when R becomes very large, P in (3-20) approaches unity and (3-19) reduces to

$$\frac{I_1(s,\omega)}{E(R,\omega)} = \left(-2j\sin\frac{\beta b}{2} \right) \frac{2\sin\dfrac{\beta s}{2}}{\beta D} \left[Z_0\cos\frac{\beta s}{2} + jZ_1\sin\frac{\beta s}{2} \right]$$

$$= \frac{\left(-2j\sin\dfrac{\beta b}{2} \right)}{\beta D} \left[Z_0\sin\beta s + jZ_1(1 - \cos\beta s) \right]$$

which is exactly the solution given by equation (2-10) for a planewave traveling in the x direction.

If the conductor spacing b is much less than the distance from the loop to the line R, and if $\beta b \ll 1$, E_x is approximately constant over the terminations and can be taken as the value at $x = b/2$. Thus the last six terms of (3-12) and of the analogous radiation solution are simplified to

$$
\frac{I_2(s,\omega)}{E(R,\omega)} =
\begin{cases}
\dfrac{bZ_0 R^2 w}{D\left[R^2+w^2\right]^{3/2}}Q_1 \\[4mm]
+\dfrac{b(Z_0\cos\beta s + jZ_1\sin\beta s)R^2(s-w)}{D\left[R^2+(s-w)^2\right]^{3/2}}Q_2 \qquad \text{Induction} \\[6mm]
\dfrac{bZ_0 Rw}{D(R^2+w^2)}Q_1 \\[4mm]
+\dfrac{b(Z_0\cos\beta s + jZ_1\sin\beta s)R(s-w)}{D\left[R^2+(s-w)^2\right]}Q_2 \qquad \text{Radiation}
\end{cases}
$$

$$(3\text{-}21)$$

where

$$
Q_1 = \cos\beta(\sqrt{R^2+w^2}-R)-j\sin\beta(\sqrt{R^2+w^2}-R)
$$

$$(3\text{-}22)$$

$$
Q_2 = \cos\beta\left(\sqrt{R^2+(s-w)^2}-R\right)-j\sin\beta\left(\sqrt{R^2+(s-w)^2}-R\right).
$$

The total current in the load due to both E_z and E_x, approximating $K(z)$ as uniform over a segment of the line of length L and E_x as constant over the terminations, is the sum of (3-19) and (3-21), that is,

$$
\frac{I(s,\omega)}{E(R,\omega)} = \frac{I_1(s,\omega)}{E(R,\omega)} + \frac{I_2(s,\omega)}{E(R,\omega)}.
$$

This solution, as opposed to (3-12), contains no integrals and thus greatly reduces the computational effort required.

3.3 NUMERICAL EXAMPLE FOR SMALL LOOP EXCITATION

A transmission line with a conductor spacing of 0.01 m and conductor diameter of 0.1 cm (representative of the dimensions of TV twin lead and two-conductor power cords) is selected as an example. The line is 10 m long, and the loop is adjacent to the middle of the line and located 2 m away. The end of the line is terminated in its characteristic impedance ($Z_2 = Z_0 = 360$). The transfer function for the current in the load Z_2 is given in Fig. 3-3 for $Z_1 = 10^7$ (open-circuited) and in Fig. 3-4 for $Z_1 = 1$ (short-circuited).

The solid curves in Figs. 3-3 and 3-4 were plotted using the approximate solution given by (3-19). [For the geometry in this example, the current due to E_x given by (3-21) is negligible compared with the current due to E_z. The currents calculated from (3-14) and (3-15) are also plotted for a number of frequencies. Agreement between the approximate solution and the more rigorous solution from (3-14) and (3-15) is excellent.

Also shown in Figs. 3-3 and 3-4 are the transfer functions for the current in the load when the line is illuminated over its entire length by a plane wave traveling in the x direction (see Section 2.2). It is seen that even though the fields from the loop effectively illuminate only a small segment of the transmission line, they excite a greater current due to their spatial nonuniformity. That is, while a plane wave induces a transmission-line mode current on the line by virtue of the phase difference of the wave on the two conductors, the current produced by a nearby source is due to both a phase difference and an amplitude difference of the fields on the two conductors.

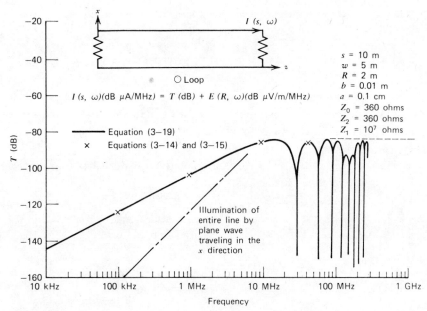

FIGURE 3-3. Current transfer function for illumination at middle of line (Z_1 open-circuited).

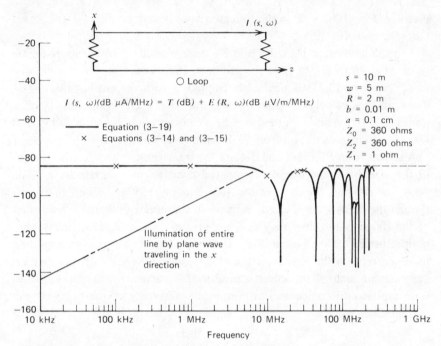

FIGURE 3-4. Current transfer function for illumination at middle of line (Z_1 short-circuited).

3.4 EXACT EQUATIONS FOR A LINE EXCITED BY A SHORT DIPOLE[†]

A short dipole antenna illuminating a two-wire transmission line is shown in Fig. 3-5. The dipole is parallel to the two conductors and in the same plane. The coordinates and geometry are the same as in Section 3.2 for the loop excitation case.

The transverse and radial field components of a short dipole are given by Kraus in reference [2] as

$$E_\theta (r,\theta,\omega) = \frac{M(\omega)\sin\theta\,\omega^2}{\varepsilon c^3}\left[\frac{j}{\beta r} + \frac{1}{(\beta r)^2} - \frac{j}{(\beta r)^3}\right]\varepsilon^{-j\beta r} \qquad (3\text{-}23)$$

$$E_r (r,\theta,\omega) = \frac{2M(\omega)\cos\theta\,\omega^2}{\varepsilon c^3}\left[\frac{1}{(\beta r)^2} - \frac{j}{(\beta r)^3}\right]\varepsilon^{-j\beta r} \qquad (3\text{-}24)$$

[†]Sections 3.4 and 3.5 are based on material from "Load Current Spectrum of a Two-Wire Transmission Line Excited by a Nearby Dipole" by A. A. Smith, Jr., Proceedings of the Electromagnetic Compatibility Symposium, Montreux, May 20–22, 1975.

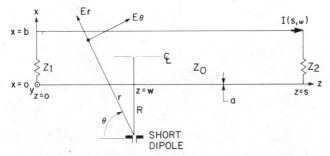

FIGURE 3-5. Geometry for transmission line and dipole.

where $M(\omega)$ is a function of the dipole driving current spectrum and dipole length.

From Fig. 3-5, we have

$$E_z^i(x,z,\omega) = E_\theta(r,\theta,\omega)\sin\theta - E_r(r,\theta,\omega)\cos\theta \qquad (3\text{-}25)$$

$$E_x^i(x,z,\omega) = E_\theta(r,\theta,\omega)\cos\theta + E_r(r,\theta,\omega)\sin\theta \qquad (3\text{-}26)$$

or

$$E_z^i(x,z,\omega) =$$

$$\frac{M(\omega)\omega^2}{\varepsilon c^3}\left[\frac{j\sin^2\theta}{\beta r} + \frac{(\sin^2\theta - 2\cos^2\theta)}{(\beta r)^2} - \frac{j(\sin^2\theta - 2\cos^2\theta)}{(\beta r)^3}\right]\varepsilon^{-j\beta r} \qquad (3\text{-}27)$$

$$E_x^i(x,z,\omega) = \frac{M(\omega)\omega^2\sin\theta\cos\theta}{\varepsilon c^3}\left[\frac{j}{(\beta r)} + \frac{3}{(\beta r)^2} - \frac{j3}{(\beta r)^3}\right]\varepsilon^{-j\beta r}. \qquad (3\text{-}28)$$

If (3-27) and (3-28) are normalized to $E(R,\omega) = E_\theta(R,\pi/2,\omega)$ and separated into their induction field ($\beta r \ll 1$) and radiation field ($\beta r \gg 1$) components, the result is

$$\frac{E_z^i(x,z,\omega)}{E(R,\omega)} = \begin{cases} \left(\dfrac{R}{r}\right)^3(\sin^2\theta - 2\cos^2\theta)\varepsilon^{-j\beta(r-R)} & \text{Induction} \\[2mm] \dfrac{R}{r}\sin^2\theta\,\varepsilon^{-j\beta(r-R)} & \text{Radiation} \end{cases} \qquad (3\text{-}29)$$

$$\frac{E_x^i(x,z,\omega)}{E(R,\omega)} = \begin{cases} 3\left(\dfrac{R}{r}\right)^3\sin\theta\cos\theta\,\varepsilon^{-j\beta(r-R)} & \text{Induction} \\[2mm] \dfrac{R}{r}\sin\theta\cos\theta\,\varepsilon^{-j\beta(r-R)} & \text{Radiation.} \end{cases} \qquad (3\text{-}30)$$

On the upper conductor of the transmission line (at $x = b$)

$$h_1(z) = \frac{R}{r} = \frac{R}{\sqrt{(w-z)^2 + (R+b/2)^2}}$$

$$h_2(z) = \sin^2\theta = \frac{(R+b/2)^2}{(w-z)^2 + (R+b/2)^2}$$

$$h_3(z) = \cos^2\theta = \frac{(w-z)^2}{(w-z)^2 + (R+b/2)^2}$$

$$h_4(z) = r - R = \sqrt{(w-z)^2 + (R+b/2)^2} - R.$$

On the lower conductor of the transmission line (at $x = 0$)

$$h_5(z) = \frac{R}{r} = \frac{R}{\sqrt{(w-z)^2 + (R-b/2)^2}}$$

$$h_6(z) = \sin^2\theta = \frac{(R-b/2)^2}{(w-z)^2 + (R-b/2)^2}$$

$$h_7(z) = \cos^2\theta = \frac{(w-z)^2}{(w-z)^2 + (R-b/2)^2}$$

$$h_8(z) = r - R = \sqrt{(w-z)^2 + (R-b/2)^2} - R.$$

On the left-hand termination (at $z = 0$)

$$h_9(x) = \frac{R}{r} = \frac{R}{\sqrt{(R-b/2+x)^2 + w^2}}$$

$$h_{10}(x) = \sin\theta = \frac{R-b/2+x}{\sqrt{(R-b/2+x)^2 + w^2}}$$

$$h_{11}(x) = \cos\theta \frac{w}{\sqrt{(R-b/2+x)^2+w^2}}$$

$$h_{12}(x) = r - R = \sqrt{(R-b/2+x)^2+w^2} - R.$$

On the right-hand termination (at $z = s$)

$$h_{13}(x) = \frac{R}{r} = \frac{R}{\sqrt{(R-b/2+x)^2+(s-w)^2}}$$

$$h_{14}(x) = \sin\theta = \frac{R-b/2+x}{\sqrt{(R-b/2+x)^2+(s-w)^2}}$$

$$h_{15}(x) = \cos\theta = \frac{-(s-w)}{\sqrt{(R-b/2+x)^2+(s-w)^2}}$$

$$h_{16}(x) = r - R = \sqrt{(R-b/2+x)^2+(s-w)^2} - R.$$

Substituting the $h(z)$ and $h(x)$ functions defined previously into (3-29) and (3-30) yields

$$\frac{E_z^i(b,z,\omega)}{E(R,\omega)} = \begin{cases} h_1^3(h_2-2h_3)(\cos\beta h_4 - j\sin\beta h_4) & \text{Induction} \\ h_1 h_2(\cos\beta h_4 - j\sin\beta h_4) & \text{Radiation} \end{cases} \qquad (3\text{-}31)$$

$$\frac{E_z^i(0,z,\omega)}{E(R,\omega)} = \begin{cases} h_5^3(h_6-2h_7)(\cos\beta h_8 - j\sin\beta h_8) & \text{Induction} \\ h_5 h_6(\cos\beta h_8 - j\sin\beta h_8) & \text{Radiation} \end{cases} \qquad (3\text{-}32)$$

$$\frac{E_x^i(x,0,\omega)}{E(R,\omega)} = \begin{cases} 3h_9^3 h_{10}h_{11}(\cos\beta h_{12} - j\sin\beta h_{12}) & \text{Induction} \\ h_9 h_{10}h_{11}(\cos\beta h_{12} - j\sin\beta h_{12}) & \text{Radiation} \end{cases} \qquad (3\text{-}33)$$

$$\frac{E_x^i(x,s,\omega)}{E(R,\omega)} = \begin{cases} 3h_{13}^3 h_{14}h_{15}(\cos\beta h_{16} - j\sin\beta h_{16}) & \text{Induction} \\ h_{13}h_{14}h_{15}(\cos\beta h_{16} - j\sin\beta h_{16}) & \text{Radiation.} \end{cases} \qquad (3\text{-}34)$$

Substituting the induction field terms from (3-31) through (3-34) into equation (3-2) in Section 3.1 yields

$$\frac{I(s,\omega)}{E(R,\omega)} = \frac{1}{D}\left[Z_0\int_0^s h_1^3(h_2-2h_3)\cos\beta z\cos\beta h_4\,dz \right.$$

$$-jZ_0\int_0^s h_1^3(h_2-2h_3)\cos\beta z\sin\beta h_4\,dz$$

$$+jZ_1\int_0^s h_1^3(h_2-2h_3)\sin\beta z\cos\beta h_4\,dz$$

$$+Z_1\int_0^s h_1^3(h_2-2h_3)\sin\beta z\sin\beta h_4\,dz$$

$$-Z_0\int_0^s h_5^3(h_6-2h_7)\cos\beta z\cos\beta h_8\,dz$$

$$+jZ_0\int_0^s h_5^3(h_6-2h_7)\cos\beta z\sin\beta h_8\,dz$$

$$-jZ_1\int_0^s h_5^3(h_6-2h_7)\sin\beta z\cos\beta h_8\,dz$$

$$-Z_1\int_0^s h_5^3(h_6-2h_7)\sin\beta z\sin\beta h_8\,dz$$

$$+Z_0\int_0^b 3h_9^3 h_{10}h_{11}\cos\beta h_{12}\,dx$$

$$-jZ_0\int_0^b 3h_9^3 h_{10}h_{11}\sin\beta h_{12}\,dx$$

$$-Z_0\cos\beta s\int_0^b 3h_{13}^3 h_{14}h_{15}\cos\beta h_{16}\,dx$$

$$+jZ_0\cos\beta s\int_0^b 3h_{13}^3 h_{14}h_{15}\sin\beta h_{16}\,dx$$

$$-jZ_1\sin\beta s\int_0^b 3h_{13}^3 h_{14}h_{15}\cos\beta h_{16}\,dx$$

$$\left. -Z_1\sin\beta s\int_0^b 3h_{13}h_{14}h_{15}\sin\beta h_{16}\,dx \right] \qquad (3\text{-}35)$$

or

$$\frac{I(s,\omega)}{E(R,\omega)} = \frac{1}{D}(G_1-jG_2+jG_3+G_4-G_5+jG_6-jG_7-G_8+G_9-jG_{10}$$

$$-G_{11}+jG_{12}-jG_{13}-G_{14}) \qquad (3\text{-}36)$$

where each G is defined by the corresponding term in (3-35).

For Z_1 and Z_2 real, the magnitude of (3-36), the induction field solution, is

$$\left| \frac{I(s,\omega)}{E(R,\omega)} \right| = \frac{1}{|D|} \left[(G_1 + G_4 - G_5 - G_8 + G_9 - G_{11} - G_{14})^2 \right.$$
$$\left. + (G_3 - G_2 + G_6 - G_7 - G_{10} + G_{12} - G_{13})^2 \right]^{1/2}. \quad (3\text{-}37)$$

Substituting the radiation field terms from (3-31) through (3-34) into (3-2) yields

$$\frac{I(s,\omega)}{E(R,\omega)} = \frac{1}{D} \left[Z_0 \int_0^s h_1 h_2 \cos \beta z \cos \beta h_4 \, dz - jZ_0 \int_0^s h_1 h_2 \cos \beta z \sin \beta h_4 \, dz \right.$$

$$+ jZ_1 \int_0^s h_1 h_2 \sin \beta z \cos \beta h_4 \, dz + Z_1 \int_0^s h_1 h_2 \sin \beta z \sin \beta h_4 \, dz$$

$$- Z_0 \int_0^s h_5 h_6 \cos \beta z \cos \beta h_8 \, dz + jZ_0 \int_0^s h_5 h_6 \cos \beta z \sin \beta h_8 \, dz$$

$$- jZ_1 \int_0^s h_5 h_6 \sin \beta z \cos \beta h_8 \, dz - Z_1 \int_0^s h_5 h_6 \sin \beta z \sin \beta h_8 \, dz$$

$$+ Z_0 \int_0^b h_9 h_{10} h_{11} \cos \beta h_{12} \, dx - jZ_0 \int_0^b h_9 h_{10} h_{11} \sin \beta h_{12} \, dx$$

$$- Z_0 \cos \beta s \int_0^b h_{13} h_{14} h_{15} \cos \beta h_{16} \, dx$$

$$+ jZ_0 \cos \beta s \int_0^b h_{13} h_{14} h_{15} \sin \beta h_{16} \, dx$$

$$- jZ_1 \sin \beta s \int_0^b h_{13} h_{14} h_{15} \cos \beta h_{16} \, dx$$

$$\left. - Z_1 \sin \beta s \int_0^b h_{13} h_{14} h_{15} \sin \beta h_{16} \, dx \right] \quad (3\text{-}38)$$

or

$$\frac{I(s,\omega)}{E(R,\omega)} = \frac{1}{D} (H_1 - jH_2 + jH_3 + H_4 - H_5 + jH_6 - jH_7 - H_8 + H_9 - jH_{10}$$
$$- H_{11} + jH_{12} - jH_{13} - H_{14})$$

where each H is defined by the corresponding term in (3-38).

The magnitude of the radiation field solution (Z_1 and Z_2 real) is

$$\left| \frac{I(s,\omega)}{E(R,\omega)} \right| = \frac{1}{|D|} \left[(H_1 + H_4 - H_5 - H_8 + H_9 - H_{11} - H_{14})^2 \right.$$
$$\left. + (H_3 - H_2 + H_6 - H_7 - H_{10} + H_{12} - H_{13})^2 \right]^{1/2}. \quad (3\text{-}39)$$

If a coupling transfer function T is defined as

$$T = 20 \log_{10} \left| \frac{I(s,\omega)}{E(R,\omega)} \right|$$

then

$$I(s,\omega) \ (\text{dB } \mu\text{A}/\text{MHz}) = E(R,\omega) \ (\text{dB } \mu\text{V}/\text{m}/\text{MHz}) + T$$

where T is found from (3-37) for the induction field and from (3-39) for the radiation field.

3.5 NUMERICAL EXAMPLE FOR DIPOLE EXCITATION

Load-current transfer functions for a line excited by a short dipole are given in Figs. 3-6 and 3-7. The load impedance Z_2 is equal to the characteristic impedance. In Fig. 3-6, $Z_1 = 10^7$ (open-circuited) and in Fig. 3-7, $Z_1 = 1$ (short-circuited). The dimensions of the transmission line and the proximity of the dipole are the same as used for the loop excitation example in Section 3.4.

The solid curves in Figs. 3-6 and 3-7 are calculated from equations (3-37) and (3-39) using a computer. The broken curves are the load-current transfer functions calculated from equation (2-10) for illumination of the line by a plane wave traveling in the y direction.

Below a few hundred kilohertz the dipole excites a significantly greater current on the line than the plane wave. This is not an obvious result, since the dipole effectively illuminates only a small segment of the line, whereas the plane wave illuminates the entire length of the line. It is, of course, a consequence of the amplitude nonuniformity of the fields from nearby sources. That is, the differential mode current excited by the nonuniform dipole field is due to both the phase difference and the amplitude difference of the wave on the two conductors, but the plane wave excites a differential mode current by virtue of a phase difference only.

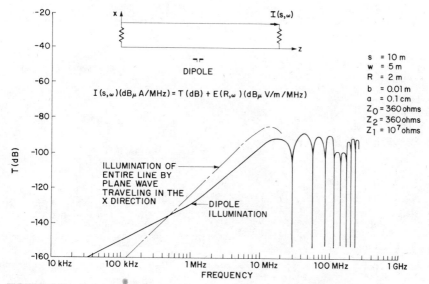

FIGURE 3-6. Current transfer function for illumination at middle of line (Z_1 open-circuited).

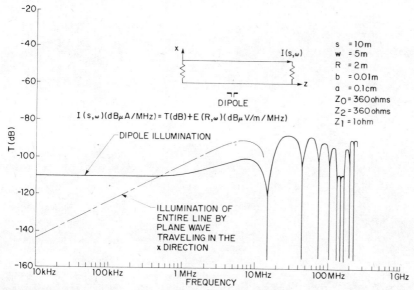

FIGURE 3-7. Current transfer function for illumination at middle of line (Z_1 short-circuited).

CHAPTER
FOUR

SHIELDED CABLES

$\mathbf{T}$he subject of this chapter is the excitation of load currents inside shielded cables by external electromagnetic fields. In Section 4.1 the theory of cable penetration is briefly reviewed. Formulas for sheath current distributions are derived in Section 4.2. In Section 4.3 the interior load current spectrum for a particular field excitation is obtained by integrating the sheath current distribution over the length of the cable.

4.1 COUPLING THEORY

The illumination of a shielded cable by an external electromagnetic field excites a current distribution on the outer shield (or sheath). Since the shield is not a perfect conductor, this current penetrates the shield and produces a voltage distribution along the inside length of the cable. This voltage distribution in turn produces a current in the interior load impedances. The pertinent geometry is shown in Fig. 4-1 where

E electric field that illuminates the cable
$I(z)$ sheath current distribution
s length of cable, meters
$b/2$ height of cable above ground plane, meters
a outside diameter of cable, meters
$Z_1/2, Z_2/2$ terminating impedances of cable shield treated as a single-wire transmission line over a ground plane
$Z_0/2$ characteristic impedance of cable shield treated as a single-wire transmission line over a ground plane
Z_a, Z_b interior load impedances
Z_c characteristic impedance of interior of cable
I_L current in the interior load impedance Z_b.

The length of the cable in Fig. 4-1 is taken to be much greater than its height above the ground plane ($s \gg b$) so that $I(z)$ is the only current of consequence. Otherwise, the contribution from $I(x)$ must be included.

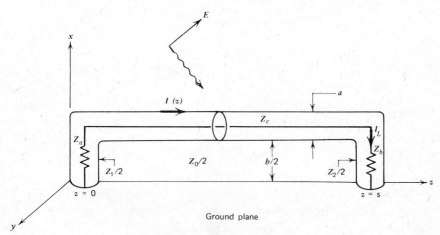

FIGURE 4-1. Shielded cable geometry.

The outer sheath current and the voltage induced along the inside of the cable are related by the surface transfer impedance. The defining relation is

$$dV(z) = Z_t I(z) dz \qquad (4\text{-}1)$$

where

Z_t surface transfer impedance, ohms/meter
V voltage induced along the inside of the cable.

(Further information on cable shielding and surface transfer impedance may be found in Refs. [1] to [4] in the Bibliography).

The current spectrum $I_L(\omega)$ in the interior load impedance Z_b is obtained by integrating (4-1) over the length of the cable. By analogy, the appropriate transmission line equation is equation (1-6) with $K(z) = Z_t I(z, \omega)$ and with $E_x = 0$. Thus

$$I_L(\omega) = \frac{Z_t}{P} \int_0^s I(z, \omega) \left[Z_c \cos \beta_i z + j Z_a \sin \beta_i z \right] dz \qquad (4\text{-}2)$$

where

$P = (Z_c Z_a + Z_c Z_b) \cos \beta_i s + j (Z_c^2 + Z_a Z_b) \sin \beta_i s$
$\omega = 2\pi f$ radian frequency
$\beta_i = 2\pi / \lambda_i$ wave number (inside cable).

If the cable is short compared to a wavelength ($\beta_i s \ll 1$), and if the average current distribution is defined as

$$I^{av}(\omega) = \int_0^s I(z, \omega)\, dz,$$

then (4-2) reduces to

$$I_L(\omega) = \frac{Z_t s I^{av}(\omega)}{Z_a + Z_b}.$$ (4-3)

Although Fig. 4-1 depicts a coaxial cable, the approach is equally valid for any shielded cable, whether balanced or unbalanced (e.g., triaxial, shielded pair, shielded multiconductor, shielded multicoax). It is necessary only to measure the appropriate surface transfer impedance.

4.2 SHEATH CURRENT DISTRIBUTIONS

The cable shield, treated as a single-wire transmission line over ground, is shown in Fig. 4-2 being illuminated by the four electromagnetic fields for which the shield current distributions are derived.

The two E_x fields (one traveling in the z direction and the other traveling in the $-y$ direction) are plane-wave fields and are representative, for instance, of AM broadcast station fields, which are vertically polarized.

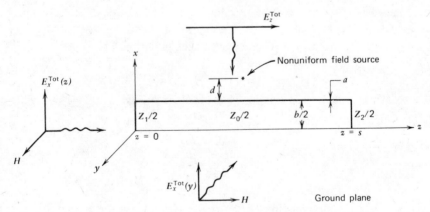

FIGURE 4-2. Cable shield over ground plane and illuminating fields.

These fields excite only the shield terminations (vertical drops).

The E_z field is also a plane wave that uniformly illuminates the entire length of the cable and is representative of TV and FM fields, which are mainly horizontally polarized.

The nonuniform field source, located at the center of the cable ($z = s/2$) a distance d away, illuminates only a small portion of the cable, and typifies those situations in which nearby or local sources of electromagnetic noise excite the cable.

It should be noted that all current distributions are in terms of the total field (incident plus ground reflected field) rather than in terms of the incident field, since, in practice only the total field can be measured. When using measured field strength values obtained at a particular height above ground to predict current distributions on cables at a different height, the field height gain must be accounted for (especially at the higher frequencies where ground wave field strength is directly proportional to height above ground). See, for instance, the height gain curves in Ref. [5].

E_x Traveling in the z Direction

The current distribution for this case is derived from the solution for a two-wire transmission line driven at both ends by the appropriate voltage generators. Denote the field, which travels in the z direction, as

$$E_x(z,\omega) = E_x^{\text{Tot}}(\omega)\exp(-j\beta z). \qquad (4\text{-}4)$$

If the phase reference is taken at $z = 0$, the fields that excite the left- and right-hand terminations in Fig. 4-2 are, respectively,

$$E_x(0,\omega) = E_x^{\text{Tot}}(\omega)$$

$$\qquad (4\text{-}5)$$

$$E_x(s,\omega) = E_x^{\text{Tot}}(\omega)\exp(-j\beta s).$$

Then the voltage generators at the ends of the line are

$$\frac{b}{2}E_x^{\text{Tot}}(\omega)$$

$$\qquad (4\text{-}6)$$

$$\frac{b}{2}E_x^{\text{Tot}}(\omega)\exp(-j\beta s).$$

The sheath current distribution, derived from the solution for a transmission line driven at the ends by the voltage generators, is

$$I(z,\omega) = \frac{bE_x^{\text{Tot}}}{D} \{(Z_0 - Z_1)\sin\beta s \sin\beta z$$

$$+ j(Z_0 + Z_2)\sin\beta s \cos\beta z - j(Z_1 + Z_2)\cos\beta s \sin\beta z\} \quad (4\text{-}7)$$

where

$$Z_0 = 120\ln(2b/a) = 276\log_{10}(2b/a)$$
$$D = (Z_0 Z_1 + Z_0 Z_2)\cos\beta s + j(Z_0^2 + Z_1 Z_2)\sin\beta s.$$

We define the sheath current distribution transfer function T_1 as the sheath current distribution in (4-7) normalized to the electric field. That is,

$$T_1 = \frac{I(z,\omega)}{E_x^{\text{Tot}}}. \quad (4\text{-}8)$$

In decibels

$$T_1\,(\text{dB}) = 20\log_{10}|T_1|. \quad (4\text{-}9)$$

Then, given T_1 and the field, the current distribution is found from

$$I(z,\omega)\,(\text{dB}\,\mu\text{A}) = E_x^{\text{Tot}}\,(\text{dB}\,\mu\text{V}) + T_1\,(\text{dB}). \quad (4\text{-}10)$$

E_x Traveling in the $-y$ Direction

Since this field arrives at both terminations simultaneously, as shown in Fig. 4-2, the sheath current distribution is derived from the solution for a two-wire transmission line driven at both ends by in-phase voltage generators of magnitude $(b/2)E_x^{\text{Tot}}$. The result is

$$I(z,\omega) = \frac{bE_x^{\text{Tot}}(\omega)}{D} \{Z_0[\cos\beta(s-z) - \cos\beta z]$$

$$+ j[Z_2\sin\beta(s-z) - Z_1\sin\beta z]\} \quad (4\text{-}11)$$

where

$$Z_0 = 120\ln(2b/a) = 276\log_{10}(2b/a)$$
$$D = (Z_0 Z_1 + Z_0 Z_2)\cos\beta s + j(Z_0^2 + Z_1 Z_2)\sin\beta s.$$

As previously, the sheath current distribution transfer function for this case is defined as

$$T_2 = \frac{I(z,\omega)}{E_x^{\text{Tot}}} \tag{4-12}$$

and, in decibels

$$T_2 \,(\text{dB}) = 20\log_{10}|T_2|. \tag{4-13}$$

Uniform E_z

The sheath current at a point $z*$ on the cable in Fig. 4-2 due to an arbitrary (nonuniform) field in the z direction is given by equation (1-10) with

$$K(z) = E_z^{\text{Tot}}(z) \quad \text{and} \quad E_x = 0.$$

Thus

$$I(z*) = \frac{Z_0 \cos \beta\,(s - z*) + jZ_2 \sin \beta\,(s - z*)}{Z_0 D}$$

$$\times \int_0^{z*} 2E_z^{\text{Tot}}(z)\left[Z_0 \cos \beta z + jZ_1 \sin \beta z \right] dz + \frac{Z_0 \cos \beta z* + jZ_1 \sin \beta z*}{Z_0 D}$$

$$\times \int_{z*}^{s} 2E_z^{\text{Tot}}(z)\left[Z_0 \cos \beta\,(s - z) + jZ_2 \sin \beta\,(s - z) \right] dz. \tag{4-14}$$

For a plane wave parallel to the length of the cable, the field in the z direction is constant, that is,

$$E_z^{\text{Tot}}(z) = E_z^{\text{Tot}}.$$

Then the sheath current distribution from (4-14) reduces to (changing the dummy variable from $z*$ to z)

$$I(z,\omega) = \frac{-j2E_z^{\text{Tot}}(\omega)}{Z_0 \beta}$$

$$\times \left\{ 1 - \frac{\left[Z_0 Z_2 \cos \beta z + Z_0 Z_1 \cos \beta\,(s - z) \right] + j\left[Z_1 Z_2 \sin \beta z + Z_1 Z_2 \sin \beta\,(s - z) \right]}{D} \right\} \tag{4-15}$$

where D and Z_0 are as previously defined.

The sheath current distribution transfer function for this case is

$$T_3 = \frac{I(z,\omega)}{E_z^{\text{Tot}}(\omega)}$$

(4-16)

and

$$T_3\,(\text{dB}) = 20\log_{10}|T_3|.$$

(4-17)

Nonuniform E_z

For the source of the nonuniform field we choose a small loop that lies in the plane of the cable. The only electric field component produced by a loop is a transverse field that has a $1/r^2$ induction field attenuation rate and a $1/r$ radiation field attenuation, where r is the radial distance from the loop. The loop is located at $z = w$ and is d meters from the cable. It is assumed that the loop is close to the cable so that the ground reflected field is negligible compared with the direct field. Figure 4-3 shows the details. The true field distribution in the z direction that illuminates the cable is approximated by a rectangular field distribution of magnitude $E_z(d)$ and width L.

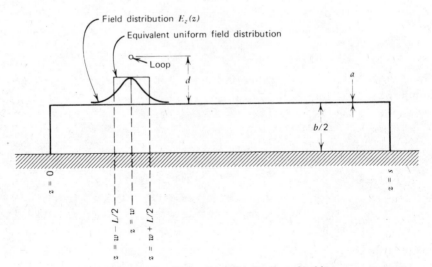

FIGURE 4-3.   Nonuniform illumination of cable.

The sheath current distribution is obtained by integrating (4-14) with

$$E_z^{Tot}(z) = \begin{cases} E_z(d) & w - L/2 < z < w + L/2 \\ 0 & \text{elsewhere} \end{cases}$$

The result is

$$I(z,\omega) = \frac{4E(d,\omega)\sin\dfrac{\beta L}{2}}{Z_0 \beta D} \left[Z_0 \cos\beta(s-z) + jZ_2 \sin\beta(s-z) \right]$$

$$\times \left[Z_0 \cos\beta w + jZ_1 \sin\beta w \right] \qquad \text{for } z > w + L/2 \qquad (4\text{-}18)$$

where

$$L = \begin{cases} 2\,d & \beta d < 1 \text{ (induction field)} \\ 3.14\,d & \beta d > 1 \text{ (radiation field)}. \end{cases}$$

The sheath current distribution transfer function is

$$T_4 = \frac{I(z,\omega)}{E_z(d)} \qquad (4\text{-}19)$$

and

$$T_4 \text{ (dB)} = 20\log_{10}|T_4|. \qquad (4\text{-}20)$$

Numerical Results

Sheath current distributions for a 100-m-long cable 5 m above ground, and with vertical drops at both ends, are given in Figs. 4-4 to 4-7 for the four illuminating fields. Figures 4-8 to 4-10 are the current distributions for the same cable with one end open (no vertical drop). The current distributions for a 300-m- long cable with vertical drops at both ends are given in Figs. 4-11 to 4-14.

The effect of reducing the height of these cables above ground to 1 m is to reduce T_1 and T_2 by 12 dB and to raise T_3 and T_4 by 2 dB.

Numerical results for a cable 1 m long and 1 cm above a ground plane are given in Figs. 4-15 to 4-19. The cable has vertical drops at both ends in Figs. 4-15 to 4-17, and is open at one end in Figs. 4-18 and 4-19.

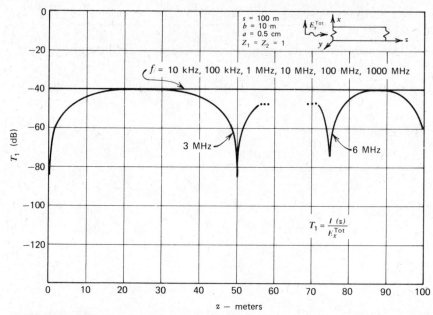

FIGURE 4-4. Sheath current distribution for 100-m cable shorted at both ends excited by uniform $E_x(z)$.

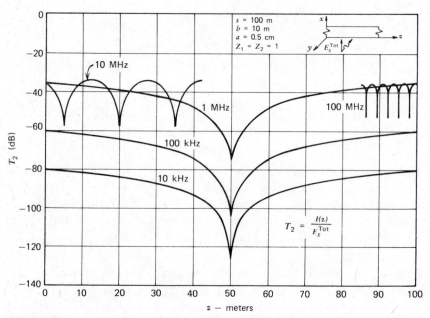

FIGURE 4-5. Sheath current distribution for 100-m cable shorted at both ends excited by uniform $E_x(y)$.

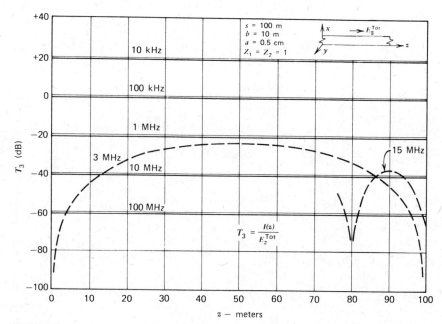

FIGURE 4-6. Sheath current distribution for 100-m cable shorted at both ends excited by uniform E_z.

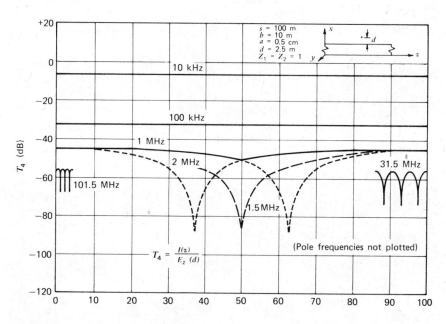

FIGURE 4-7. Sheath current distribution for 100-m cable shorted at both ends excited by nonuniform field.

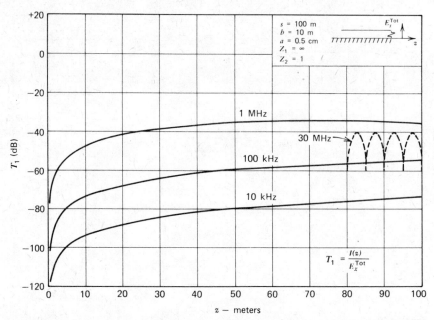

FIGURE 4-8. Sheath current distribution for 100-m cable open at one end excited by uniform E_x.

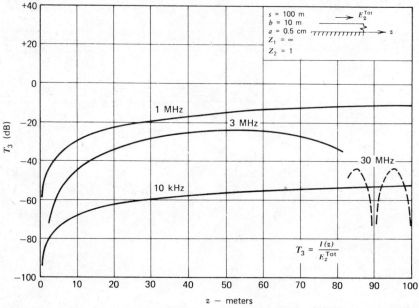

FIGURE 4-9. Sheath current distribution for 100-m cable open at one end excited by uniform E_z.

80

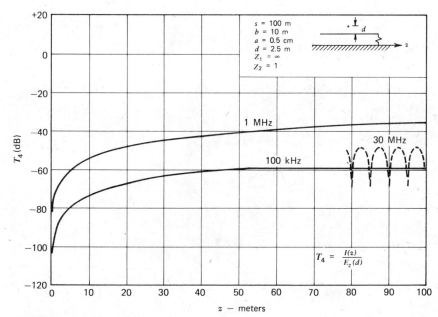

FIGURE 4-10. Sheath current distribution for 100-m cable open at one end excited at center by nonuniform field.

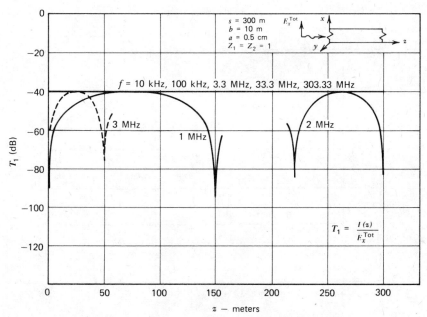

FIGURE 4-11. Sheath current distribution for 300-m cable shorted at both ends excited by uniform $E_x(z)$.

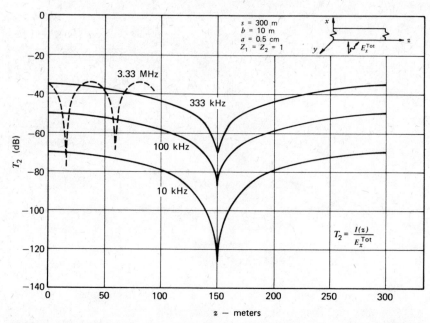

FIGURE 4-12. Sheath current distribution for 300-m cable shorted at both ends excited by $E_x(y)$.

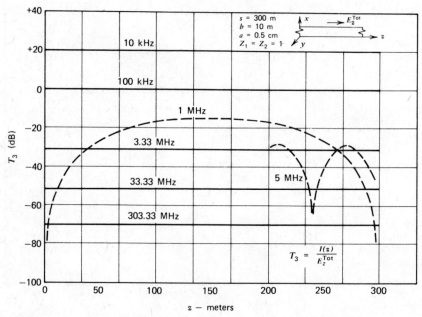

FIGURE 4-13. Sheath current distribution for 300-m cable shorted at both ends excited by uniform E_z.

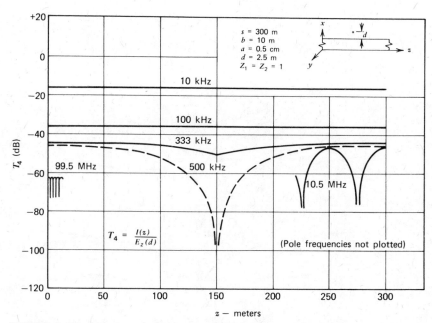

FIGURE 4-14. Sheath current distribution for 300-m cable shorted at both ends excited at the center by nonuniform field.

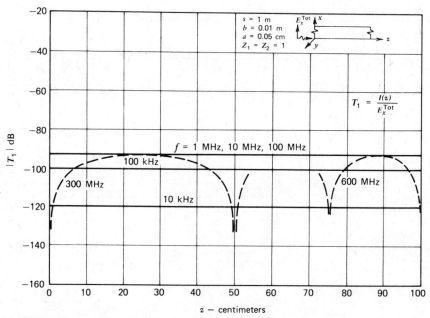

FIGURE 4-15. Sheath current distribution for 1-m cable shorted at both ends excited by uniform $E_x(z)$.

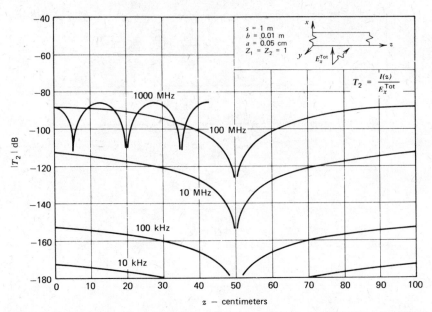

FIGURE 4-16. Sheath current distribution for 1-m cable shorted at both ends excited by uniform $E_x(y)$.

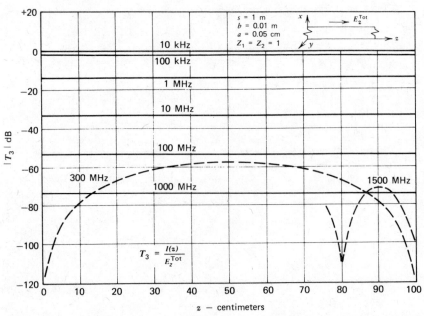

FIGURE 4-17. Sheath current distribution for 1-m cable shorted at both ends excited by uniform E_z.

84

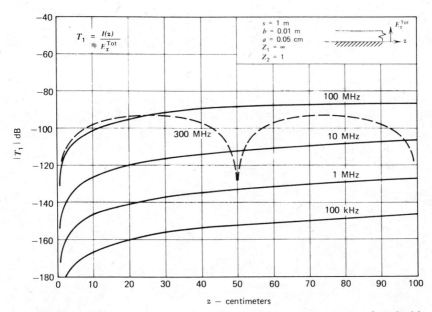

FIGURE 4-18. Sheath current distribution for 1-m cable open at one end excited by uniform E_x.

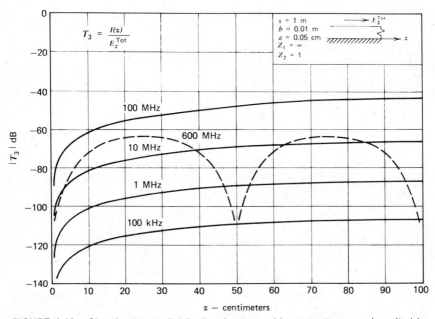

FIGURE 4-19. Sheath current distribution for 1-m cable open at one end excited by uniform E_z.

4.3 LOAD CURRENT SPECTRUM

The exact load current spectrum for a particular illuminating field is found by substituting the current distribution given by (4-7), (4-11), (4-15), or (4-18) into equation (4-2) and performing the integration. Since the resulting expressions are complicated and difficult to evaluate, this calculation is carried out only for the E_x field traveling in the z direction and with the following simplifying assumptions:

1. The interior load impedances are matched to the characteristic impedance of the cable, that is,

$$Z_a = Z_b = Z_c.$$

2. The dielectric inside the cable is air, that is, $\beta_i = \beta$.

Substituting the sheath current distribution given by (4-7) into equation (4-2) using these assumptions yields

$$I_L = \frac{Z_T b E_x^{\text{Tot}}(\omega)}{2 Z_c D} \frac{1}{\cos \beta s + j \sin \beta s}$$

$$\times \left\{ \int_0^s (Z_0 - Z_1) \sin \beta s \sin \beta z \cos \beta z \, dz \right.$$

$$+ j \int_0^s (Z_0 + Z_2) \sin \beta s \cos^2 \beta z \, dz$$

$$- j \int_0^s (Z_1 + Z_2) \cos \beta s \sin \beta z \cos \beta z \, dz$$

$$+ j \int_0^s (Z_0 - Z_1) \sin \beta s \sin^2 \beta z \, dz$$

$$- \int_0^s (Z_0 + Z_2) \sin \beta s \cos \beta z \sin \beta z \, dz$$

$$\left. + \int_0^s (Z_1 + Z_2) \cos \beta s \sin^2 \beta z \, dz \right\}.$$

After combining the first and fifth terms in the braces, we have

$$I_L = \frac{Z_T b E_x^{Tot}(\omega)}{2 Z_c D} \frac{1}{\cos \beta s + j \sin \beta s}$$

$$\times \left\{ - \int_0^s (Z_1 + Z_2) \sin \beta s \sin \beta z \cos \beta z \, dz \right.$$

$$+ \int_0^s (Z_1 + Z_2) \cos \beta s \sin^2 \beta z \, dz$$

$$+ j \int_0^s (Z_0 + Z_2) \sin \beta s \cos^2 \beta z \, dz$$

$$- j \int_0^s (Z_1 + Z_2) \cos \beta s \sin \beta z \cos \beta z \, dz$$

$$\left. + j \int_0^s (Z_0 - Z_1) \sin \beta s \sin^2 \beta z \, dz \right\}.$$

Performing the integration results in

$$I_L = \frac{Z_T b E_x^{Tot}(\omega)}{4 \beta Z_c D} \frac{1}{\cos \beta s + j \sin \beta s}$$

$$\times \left\{ -(Z_1 + Z_2) \sin \beta s \sin^2 \beta s + (Z_1 + Z_2) \cos \beta s (\beta s - \cos \beta s \sin \beta s) \right.$$

$$+ j(Z_0 + Z_2) \sin \beta s (\beta s + \sin \beta s \cos \beta s) - j(Z_1 + Z_2) \cos \beta s \sin^2 \beta s$$

$$\left. + j(Z_0 - Z_1) \sin \beta s (\beta s - \cos \beta s \sin \beta s) \right\}.$$

The final expression for the load current spectrum is obtained after expanding and collecting the terms in the braces. The result is

$$I_L(\omega) = Z_t E_x^{Tot}(\omega)$$

$$\times \left\{ \frac{b \left[(Z_1 + Z_2)(\beta s \cos \beta s - \sin \beta s) + j(2Z_0 + Z_2 - Z_1) \beta s \sin \beta s \right]}{4 \beta Z_c D (\cos \beta s + j \sin \beta s)} \right\}.$$

$$(4\text{-}21)$$

This load current spectrum, normalized to the surface transfer impedance Z_t and the electric field E_x, is plotted in Figs. 4-20 to 4-22. Figure 4-20 is for 100-m- and 300-m-long cables with vertical drops at both ends. Figure 4-21 is for the 100-m cable open at one end. Figure 4-22 is for a 1-m-long cable with and without vertical drops at one end.

Also plotted in Figs. 4-20 to 4-22 for comparison is the approximate solution given by equation (4-3), also normalized to the surface transfer impedance and E field. That is, from (4-3) and (4-8)

$$I_L(\omega) = \frac{Z_t s I^{av}(\omega)}{Z_a + Z_b} = \frac{Z_t s T_1^{av} E_x^{Tot}}{Z_a + Z_b}$$

or

$$\frac{I_L(\omega)}{Z_t E_X^{Tot}} = \frac{s T_1^{av}}{Z_a + Z_b} \tag{4-22}$$

where T_1^{av} is the average current distribution transfer function obtained from Figs. 4-4, 4-8, 4-11, 4-15, and 4-18.

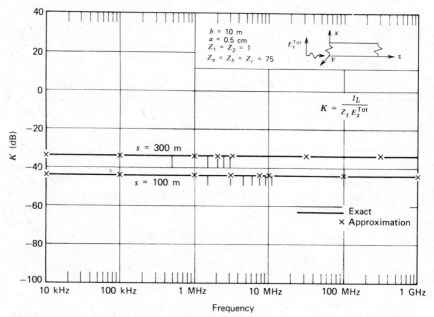

FIGURE 4-20. Load current spectrum, normalized to Z_t, for 100-m and 300-m cables excited by uniform $E_x(z)$ (sheath shorted at both ends).

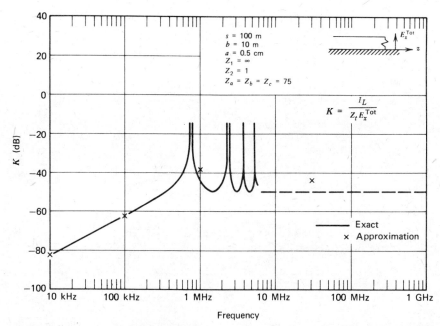

FIGURE 4-21. Load current spectrum, normalized to Z_t, for 100-m cable excited by uniform E_x (sheath open at one end).

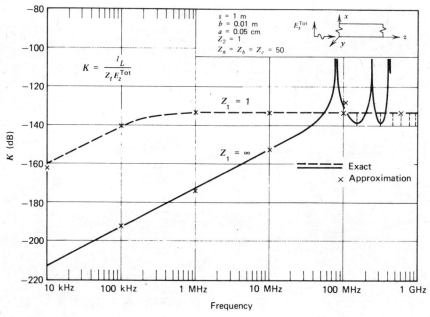

FIGURE 4-22. Load current spectrum, normalized to Z_t, for 1-m cable excited by uniform $E_x(z)$.

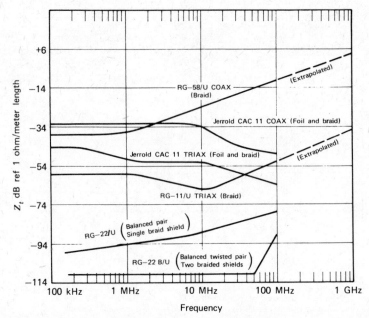

FIGURE 4-23. Measured transfer impedances of selected shielded cables (due to Shah).

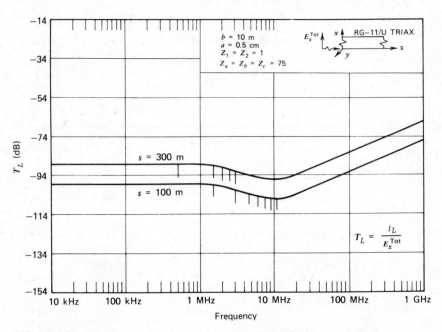

FIGURE 4-24. Load current spectrum for 100-m and 300-m lengths of RG-11/U triax excited by uniform $E_x(z)$ (sheath shorted at both ends).

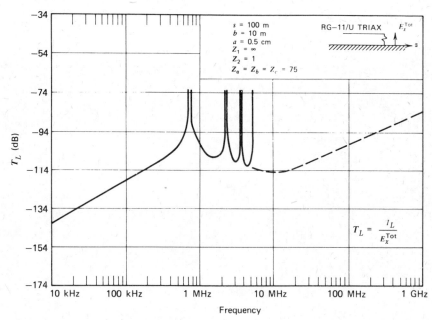

FIGURE 4-25. Load current spectrum for 100-m length of RG-11/U triax excited by uniform E_x (sheath open at one end).

As noted in Figs. 4-20 to 4-22, the load current obtained by using the simple approximate formula of equation (4-22) and the average current distribution on the cable sheath is very close to the exact current calculated from equation (4-21).

The surface transfer impedances of a number of shielded cables, measured by Dr. Arvind Shah of IBM, are given in Fig. 4-23.

Figures 4-24 and 4-25 are the load current spectrums for RG-11/U Triax obtained from equation (4-21) and the surface transfer impedance in Fig. 4-23.

A

CHARACTERISTIC IMPEDANCE GRAPH

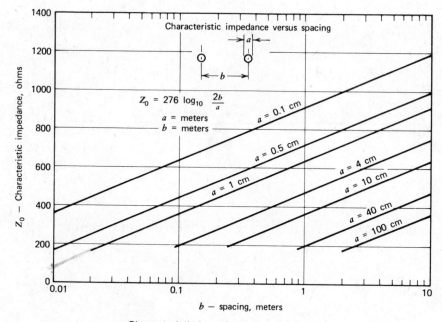

Characteristic impedance versus spacing.

B

SOME USEFUL IDENTITIES

TRANSMISSION LINES AND FIELDS

Relations that are useful in deriving and transforming transmission line and field equations are given below following the definition of symbols.

β wave number or phase constant

λ wavelength

λ_0 free space wavelength

$\omega = 2\pi f$ radian frequency

f frequency, hertz

f_{MHz} frequency, megahertz

v velocity of propagation

$c = 1/\sqrt{\mu_0 \varepsilon_0}$ velocity of propagation in free space, 3×10^8 m/sec

μ permeability

μ_0 permeability of free space, $4\pi \times 10^{-7}$ henrys/meter

ε permittivity

ε_0 permittivity of free space, 8.854×10^{-12} farads/meter

E electric field strength, volts/meter

H magnetic field strength, amperes/meter

$B = \mu H$ magnetic flux density, webers/meter^2

η intrinsic impedance of medium

$\eta_0 = (\mu_0/\varepsilon_0)^{1/2} = 120\pi$ intrinsic impedance of free space.

In any medium

$$\beta = \frac{2\pi}{\lambda} = \frac{2\pi f}{v} = \frac{\omega}{v}$$

and in free space

$$\beta_0 = \frac{2\pi}{\lambda_0} = \frac{2\pi f}{c} = \frac{\pi f_{\mathrm{MHz}}}{150} = \frac{\omega}{c}.$$

Similarly, in any medium

$$\frac{\beta}{\varepsilon \eta \omega} = \frac{1}{\varepsilon \eta v} = 1$$

and

$$\frac{\eta \beta}{\omega \mu} = \frac{\eta}{\mu v} = 1.$$

96

In free space

$$\frac{\beta_0}{120\pi\omega\varepsilon_0} = \frac{1}{120\pi\varepsilon_0 c} = 1$$

and

$$\frac{120\pi\beta_0}{\omega\mu_0} = \frac{120\pi}{\mu_0 c} = 1.$$

For a uniform plane wave in free space

$$E = 120\pi H = \left(\frac{\mu_0}{\varepsilon_0}\right)^{1/2} H = \frac{B}{(\mu_0\varepsilon_0)^{1/2}} = cB = \frac{\omega}{\beta_0} B.$$

HYPERBOLIC FUNCTIONS

$$\sinh x = \frac{\varepsilon^x - \varepsilon^{-x}}{2}$$

$$\cosh x = \frac{\varepsilon^x + \varepsilon^{-x}}{2}$$

$$\tanh x = \frac{\varepsilon^x - \varepsilon^{-x}}{\varepsilon^x + \varepsilon^{-x}}$$

$$\varepsilon^x = \cosh x + \sinh x$$

$$\varepsilon^{-x} = \cosh x - \sinh x$$

$$\sinh x = -\sinh(-x)$$

$$\cosh x = \cosh(-x)$$

$$\tanh x = -\tanh(-x)$$

$$\tanh x = \frac{\sinh x}{\cosh x}$$

$$\cosh^2 x - \sinh^2 x = 1$$

$$\sinh(x \pm y) = \sinh x \cosh y \pm \cosh x \sinh y$$

$$\cosh(x \pm y) = \cosh x \cosh y \pm \sinh x \sinh y$$

$$\tanh(x \pm y) = \frac{\tanh x \pm \tanh y}{1 \pm \tanh x \tanh y}$$

$$\sinh x + \sinh y = 2 \sinh \frac{x+y}{2} \cosh \frac{x-y}{2}$$

$$\sinh x - \sinh y = 2 \cosh \frac{x+y}{2} \sinh \frac{x-y}{2}$$

$$\cosh x + \cosh y = 2 \cosh \frac{x+y}{2} \cosh \frac{x-y}{2}$$

$$\cosh x - \cosh y = 2 \sinh \frac{x+y}{2} \sinh \frac{x-y}{2}$$

$$\sinh x + \cosh x = \frac{1 + \tanh(x/2)}{1 - \tanh(x/2)}$$

$$\sinh \frac{x}{2} = \sqrt{(\cosh x - 1)/2}$$

$$\cosh \frac{x}{2} = \sqrt{(\cosh x + 1)/2}$$

$$\tanh \frac{x}{2} = \frac{\cosh x - 1}{\sinh x} = \frac{\sinh x}{\cosh x + 1}$$

$$\sinh 2x = 2 \sinh x \cosh x$$

$$\cosh 2x = \cosh^2 x + \sinh^2 x$$

$$\tanh 2x = \frac{2 \tanh x}{1 + \tanh^2 x}$$

$$\frac{d}{dx} \sinh x = \cosh x$$

$$\frac{d}{dx} \cosh x = \sinh x$$

$$\frac{d}{dx} \tanh x = \frac{1}{\cosh^2 x}$$

$$\int \sinh x \, dx = \cosh x$$

$$\int \cosh x \, dx = \sinh x$$

$$\int \tanh x \, dx = \ln \cosh x$$

TRIGONOMETRIC FUNCTIONS

$$\sin x = \frac{\epsilon^{jx} - \epsilon^{-jx}}{2j}$$

$$\cos x = \frac{\epsilon^{jx} + \epsilon^{-jx}}{2}$$

$$\tan x = \frac{\epsilon^{jx} - \epsilon^{-jx}}{j\epsilon^{jx} + j\epsilon^{-jx}}$$

Euler's Equations

$$\epsilon^{jx} = \cos x + j \sin x$$
$$\epsilon^{-jx} = \cos x - j \sin x$$
$$\sin^2 x + \cos^2 x = 1$$
$$\sin(x \pm y) = \sin x \cos y \pm \cos x \sin y$$
$$\cos(x \pm y) = \cos x \cos y \mp \sin x \sin y$$
$$\sin x \pm \sin y = 2 \sin \frac{x \pm y}{2} \cos \frac{x \mp y}{2}$$
$$\cos x + \cos y = 2 \cos \frac{x+y}{2} \cos \frac{x-y}{2}$$
$$\cos x - \cos y = -2 \sin \frac{x+y}{2} \sin \frac{x-y}{2}$$
$$\sin x \cos y$$
$$= \frac{1}{2} \left[\sin(x+y) + \sin(x-y) \right]$$
$$\sin x \sin y$$
$$= \frac{1}{2} \left[\cos(x-y) - \cos(x+y) \right]$$

$$\cos x \sin y$$
$$= \frac{1}{2} \left[\sin(x+y) - \sin(x-y) \right]$$
$$\cos x \cos y$$
$$= \frac{1}{2} \left[\cos(x+y) + \cos(x-y) \right]$$
$$\sin 2x = 2 \sin x \cos x$$
$$\cos 2x = \cos^2 x - \sin^2 x = 1 - 2 \sin^2 x$$
$$= 2 \cos^2 x - 1$$
$$\tan 2x = 2 \tan x / 1 - \tan^2 x$$
$$\sin^2 x = \frac{1 - \cos 2x}{2}$$
$$\cos^2 x = \frac{1 + \cos 2x}{2}$$
$$\sin^2 x \cos^2 x = \frac{\sin^2 2x}{4}$$

RELATIONS BETWEEN HYPERBOLIC AND TRIGONOMETRIC FUNCTIONS

$$\sinh jx = j \sin x$$
$$\cosh jx = \cos x$$
$$\tanh jx = j \tan x$$
$$\cos jx = \cosh x$$

$$\sin jx = j \sinh x$$
$$\tan jx = j \tanh x$$
$$\sinh(u \pm jv) = \sinh u \cos v \pm j \cosh u \sin v$$
$$\cosh(u \pm jv) = \cosh u \cos v \pm j \sinh u \sin v$$

C

TABLE OF TRIGONOMETRIC FUNCTIONS

X		SIN X	COS X	TAN X	COT X	1- COS X
DEGREES	RADIANS					
0	.000	.00000	1.00000	.00000	∞	1.00000
1	.017	.01745	.99985	.01746	57.28996	.98255
2	.035	.03490	.99939	.03492	28.63625	.96510
3	.052	.05234	.99863	.05241	19.08114	.94766
4	.070	.06976	.99756	.06993	14.30067	.93024
5	.087	.08716	.99619	.08749	11.43005	.91284
6	.105	.10453	.99452	.10510	9.51436	.89547
7	.122	.12187	.99255	.12278	8.14435	.87813
8	.140	.13917	.99027	.14054	7.11537	.86083
9	.157	.15643	.98769	.15838	6.31375	.84357
10	.175	.17365	.98481	.17633	5.67128	.82635
11	.192	.19081	.98163	.19438	5.14455	.80919
12	.209	.207 1	.97815	.21256	4.70463	.79209
13	.227	.22495	.97437	.23087	4.33148	.77505
14	.244	.24192	.97030	.24933	4.01078	.75808
15	.262	.25882	.96593	.26795	3.73205	.74118
16	.279	.27564	.96126	.28675	3.48741	.72436
17	.297	.29237	.95630	.30573	3.27085	.70763
18	.314	.30902	.95106	.32492	3.07768	.69098
19	.332	.32557	.94552	.34433	2.90421	.67443
20	.349	.34202	.93969	.36397	2.74748	.65798
21	.367	.35837	.93358	.38386	2.60509	.64163
22	.384	.37461	.92718	.40403	2.47509	.62539
23	.401	.39073	.92050	.42447	2.35585	.60927
24	.419	.40674	.91355	.44523	2.24604	.59326
25	.436	.42262	.90631	.46631	2.14451	.57738
26	.454	.43837	.89879	.48773	2.05030	.56163
27	.471	.45399	.89101	.50953	1.96261	.54601
28	.489	.46947	.88295	.53171	1.88073	.53053
29	.506	.48481	.87462	.55431	1.80405	.51519
30	.524	.50000	.86603	.57735	1.73205	.50000
31	.541	.51504	.85717	.60086	1.66428	.48496
32	.559	.52992	.84805	.62487	1.60033	.47008
33	.576	.54464	.83867	.64941	1.53986	.45536
34	.593	.55919	.82904	.67451	1.48256	.44081
35	.611	.57358	.81915	.70021	1.42815	.42642
36	.628	.58779	.80902	.72654	1.37638	.41221
37	.646	.60182	.79864	.75355	1.32704	.39818
38	.663	.61566	.78801	.78129	1.27994	.38434
39	.681	.62932	.77715	.80978	1.23490	.37068
40	.698	.64279	.76604	.83910	1.19175	.35721
41	.716	.65606	.75471	.86929	1.15037	.34394
42	.733	.66913	.74314	.90040	1.11061	.33087
43	.750	.68200	.73135	.93252	1.07237	.31800
44	.768	.69466	.71934	.96569	1.03553	.30534
45	.785	.70711	.70711	1.00000	1.00000	.29289

X		SIN X	COS X	TAN X	COT X	1- COS X
DEGREES	RADIANS					
46	.803	.71934	.69466	1.03553	.96569	.28066
47	.820	.73135	.68200	1.07237	.93252	.26865
48	.838	.74314	.66913	1.11061	.90040	.25686
49	.855	.75471	.65606	1.15037	.86929	.24529
50	.873	.76604	.64279	1.19175	.83910	.23396
51	.890	.77715	.62932	1.23490	.80978	.22285
52	.908	.78801	.61566	1.27994	.78129	.21199
53	.925	.79864	.60182	1.32704	.75355	.20136
54	.942	.80902	.58779	1.37638	.72654	.19098
55	.960	.81915	.57358	1.42815	.70021	.18085
56	.977	.82904	.55919	1.48256	.67451	.17096
57	.995	.83867	.54464	1.53986	.64941	.16133
58	1.012	.84805	.52992	1.60033	.62487	.15195
59	1.030	.85717	.51504	1.66428	.60086	.14283
60	1.047	.86603	.50000	1.73205	.57735	.13397
61	1.065	.87462	.48481	1.80405	.55431	.12538
62	1.082	.88295	.46947	1.88073	.53171	.11705
63	1.100	.89101	.45399	1.96261	.50953	.10899
64	1.117	.89879	.43837	2.05030	.48773	.10121
65	1.134	.90631	.42262	2.14451	.46631	.09369
66	1.152	.91355	.40674	2.24604	.44523	.08645
67	1.169	.92050	.39073	2.35585	.42447	.07950
68	1.187	.92718	.37461	2.47509	.40403	.07282
69	1.204	.93358	.35837	2.60509	.38386	.06642
70	1.222	.93969	.34202	2.74748	.36397	.06031
71	1.239	.94552	.32557	2.90421	.34433	.05448
72	1.257	.95106	.30902	3.07768	.32492	.04894
73	1.274	.95630	.29237	3.27085	.30573	.04370
74	1.292	.96126	.27564	3.48741	.28675	.03874
75	1.309	.96593	.25882	3.73205	.26795	.03407
76	1.326	.97030	.24192	4.01078	.24933	.02970
77	1.344	.97437	.22495	4.33148	.23087	.02563
78	1.361	.97815	.20791	4.70463	.21256	.02185
79	1.379	.98163	.19081	5.14455	.19438	.01837
80	1.396	.98481	.17365	5.67128	.17633	.01519
81	1.414	.98769	.15643	6.31375	.15838	.01231
82	1.431	.99027	.13917	7.11537	.14054	.00973
83	1.449	.99255	.12187	8.14435	.12278	.00745
84	1.466	.99452	.10453	9.51436	.10510	.00548
85	1.484	.99619	.08716	11.43005	.08749	.00381
86	1.501	.99756	.06976	14.30067	.06993	.00244
87	1.518	.99863	.05234	19.08114	.05241	.00137
88	1.536	.99939	.03490	28.63625	.03492	.00061
89	1.553	.99985	.01745	57.28996	.01746	.00015
90	1.571	1.00000	.00000	∞	.00000	.00000

TABLE OF HYPERBOLIC FUNCTIONS

X	SINH X	COSH X	TANH X	X	SINH X	COSH X	TANH X
0	.00000	1.00000	.00000	0.5	.52110	1.12763	.46212
0.01	.01000	1.00005	.01000	0.51	.53240	1.13289	.46995
0.02	.02000	1.00020	.02000	0.52	.54375	1.13827	.47770
0.03	.03000	1.00045	.02999	0.53	.55516	1.14377	.48538
0.04	.04001	1.00080	.03998	0.54	.56663	1.14938	.49299
0.05	.05002	1.00125	.04996	0.55	.57815	1.15510	.50052
0.06	.06004	1.00180	.05993	0.56	.58973	1.16094	.50798
0.07	.07006	1.00245	.06989	0.57	.60137	1.16690	.51536
0.08	.08009	1.00320	.07983	0.58	.61307	1.17297	.52267
0.09	.09012	1.00405	.08976	0.59	.62483	1.17916	.52990
0.1	.10017	1.00500	.09967	0.6	.63665	1.18547	.53705
0.11	.11022	1.00606	.10956	0.61	.64854	1.19189	.54413
0.12	.12029	1.00721	.11943	0.62	.66049	1.19844	.55113
0.13	.13037	1.00846	.12927	0.63	.67251	1.20510	.55805
0.14	.14046	1.00982	.13909	0.64	.68459	1.21189	.56490
0.15	.15056	1.01127	.14889	0.65	.69675	1.21879	.57167
0.16	.16068	1.01283	.15865	0.66	.70897	1.22582	.57836
0.17	.17082	1.01448	.16838	0.67	.72126	1.23297	.58498
0.18	.18097	1.01624	.17808	0.68	.73363	1.24025	.59152
0.19	.19115	1.01810	.18775	0.69	.74607	1.24765	.59798
0.2	.20134	1.02007	.19738	0.7	.75858	1.25517	.60437
0.21	.21155	1.02213	.20697	0.71	.77117	1.26282	.61068
0.22	.22178	1.02430	.21652	0.72	.78384	1.27059	.61691
0.23	.23203	1.02657	.22603	0.73	.79659	1.27849	.62307
0.24	.24231	1.02894	.23550	0.74	.80941	1.28652	.62915
0.25	.25261	1.03141	.24492	0.75	.82232	1.29468	.63515
0.26	.26294	1.03399	.25430	0.76	.83530	1.30297	.64108
0.27	.27329	1.03667	.26362	0.77	.84838	1.31139	.64693
0.28	.28367	1.03946	.27291	0.78	.86153	1.31994	.65271
0.29	.29408	1.04235	.28213	0.79	.87478	1.32862	.65841
0.3	.30452	1.04534	.29131	0.8	.88811	1.33743	.66404
0.31	.31499	1.04844	.30044	0.81	.90152	1.34638	.66959
0.32	.32549	1.05164	.30951	0.82	.91503	1.35547	.67507
0.33	.33602	1.05495	.31852	0.83	.92863	1.36468	.68048
0.34	.34659	1.05836	.32748	0.84	.94233	1.37404	.68581
0.35	.35719	1.06188	.33638	0.85	.95612	1.38353	.69107
0.36	.36783	1.06550	.34521	0.86	.97000	1.39316	.69626
0.37	.37850	1.06923	.35399	0.87	.98398	1.40293	.70137
0.38	.38921	1.07307	.36271	0.88	.99806	1.41284	.70642
0.39	.39996	1.07702	.37136	0.89	1.01224	1.42289	.71139
0.4	.41075	1.08107	.37995	0.9	1.02652	1.43309	.71630
0.41	.42158	1.08523	.38847	0.91	1.04090	1.44342	.72113
0.42	.43246	1.08950	.39693	0.92	1.05539	1.45390	.72590
0.43	.44337	1.09388	.40532	0.93	1.06998	1.46453	.73059
0.44	.45434	1.09837	.41364	0.94	1.08468	1.47530	.73522
0.45	.46534	1.10297	.42190	0.95	1.09948	1.48623	.73978
0.46	.47640	1.10768	.43008	0.96	1.11440	1.49729	.74428
0.47	.48750	1.11250	.43820	0.97	1.12943	1.50851	.74870
0.48	.49865	1.11743	.44624	0.98	1.14457	1.51988	.75307
0.49	.50984	1.12247	.45422	0.99	1.15983	1.53141	.75736

X	SINH X	COSH X	TANH X		X	SINH X	COSH X	TANH X
1	1.17520	1.54308	.76159		4	27.28992	27.30823	.99933
1.1	1.33565	1.66852	.80050		4.1	30.16186	30.17843	.99945
1.2	1.50946	1.81066	.83365		4.2	33.33567	33.35066	.99955
1.3	1.69838	1.97091	.86172		4.3	36.84311	36.85668	.99963
1.4	1.90430	2.15090	.88535		4.4	40.71930	40.73157	.99970
1.5	2.12928	2.35241	.90515		4.5	45.00301	45.01412	.99975
1.6	2.37557	2.57746	.92167		4.6	49.73713	49.74718	.99980
1.7	2.64563	2.82832	.93541		4.7	54.96904	54.97813	.99983
1.8	2.94217	3.10747	.94681		4.8	60.75109	60.75932	.99986
1.9	3.26816	3.41773	.95624		4.9	67.14117	67.14861	.99989
2	3.62686	3.76220	.96403		5	74.20321	74.20995	.99991
2.1	4.02186	4.14431	.97045		5.1	82.00791	82.01400	.99993
2.2	4.45711	4.56791	.97574		5.2	90.63336	90.63888	.99994
2.3	4.93696	5.03722	.98010		5.3	100.16591	100.17090	.99995
2.4	5.46623	5.55695	.98367		5.4	110.70095	110.70547	.99996
2.5	6.05020	6.13229	.98661		5.5	122.34392	122.34801	.99997
2.6	6.69473	6.76901	.98903		5.6	135.21135	135.21505	.99997
2.7	7.40626	7.47347	.99101		5.7	149.43203	149.43537	.99998
2.8	8.19192	8.25273	.99263		5.8	165.14827	165.15129	.99998
2.9	9.05956	9.11458	.99396		5.9	182.51736	182.52010	.99998
3	10.01787	10.06766	.99505		6	201.71316	201.71564	.99999
3.1	11.07645	11.12150	.99595		6.1	222.92776	222.93001	.99999
3.2	12.24588	12.28665	.99668		6.2	246.37351	246.37554	.99999
3.3	13.53788	13.57476	.99728		6.3	272.28504	272.28687	.99999
3.4	14.96536	14.99874	.99777		6.4	300.92169	300.92335	.99999
3.5	16.54263	16.57282	.99818		6.5	332.57006	332.57157	1.00000
3.6	18.28546	18.31278	.99851		6.6	367.54691	367.54827	1.00000
3.7	20.21129	20.23601	.99878		6.7	406.20230	406.20353	1.00000
3.8	22.33941	22.36178	.99900		6.8	448.92309	448.92420	1.00000
3.9	24.69110	24.71135	.99918		6.9	496.13685	496.13786	1.00000

TRANSMISSION LINE THEORY

1. INTRODUCTION

Classical transmission line theory is concerned with the propagation of transverse electromagnetic (TEM) waves on uniform two-conductor lines. A variety of transmission lines fit this description, and some of the more common configurations are shown in Fig. 1. (The shielded strip-line is considered a two-wire line if the source is connected between the center strip and both outside strips. The same applies to the shielded-pair line if the source is connected between the two conductors. Obviously, other arrangements are possible. For example, sources connected between each conductor and the shield of the shielded pair, or between each shield and the center strip in the case of the shielded stripline. These lines would then be classified as multiconductor lines).

The "uniform" property of a transmission line refers to the constancy of the conductor geometry (spacing and cross-sectional area), conductor material, and the surrounding dielectric medium, over the length of the line. (The familiar tapered transmission line used for impedance matching is an example of a nonuniform line).

When the conductor spacing is much less than a quarter wavelength, the only mode that will propagate on a line is the TEM mode, which is also referred to as the "principal mode." The TEM wave is characterized by

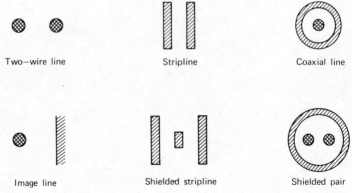

| Two—wire line | Stripline | Coaxial line |

| Image line | Shielded stripline | Shielded pair |

FIGURE 1. Common transmission lines.

electric and magnetic fields that are perpendicular to each other and to the direction of propagation. When the conductor spacing is comparable to a quarter wavelength, TE (transverse electric) and TM (transverse magnetic) modes, which are not amendable to analysis by transmission-line methods, can be supported on the line.

A short section of a two-conductor transmission line is represented schematically in terms of the distributed line constants as shown in Fig. 2. R is the series resistance per unit length, L is the series inductance per unit length, G is the shunt conductance per unit length, and C is the shunt capacitance per unit length. The resistance of a section of line of length Δz is $R\Delta z$, and so on for the other line constants.

The differential equations that describe the voltage and current along the line in the time domain are

$$-\frac{\partial v}{\partial z}\,\Delta z = (R\,\Delta z)i + (L\,\Delta z)\frac{\partial i}{\partial z}$$

$$-\frac{\partial i}{\partial z}\,\Delta z = (G\,\Delta z)v + (C\,\Delta z)\frac{\partial v}{\partial z}$$

or

$$-\frac{\partial v}{\partial z} = Ri + L\frac{\partial i}{\partial t}$$

$$-\frac{\partial i}{\partial z} = Gv + C\frac{\partial v}{\partial t}$$

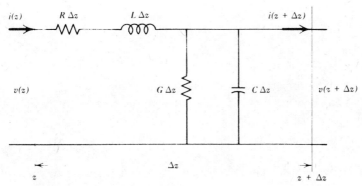

FIGURE 2.   Schematic representation of a short section of line.

where v and i are functions of both the space variable z and the time variable t, that is, $v = v(z, t)$ and $i = i(z, t)$.

In the frequency domain, since the Fourier transform of $df(t)/dt$ is $j\omega F(\omega)$, we have

$$-\frac{dV}{dz} = (R + j\omega L) I$$

$$-\frac{dI}{dz} = (G + j\omega C) V$$

where $V = V(z, \omega)$ and $I = I(z, \omega)$. The series impedance per unit length Z and shunt admittance per unit length Y are defined as

$$Z = R + j\omega L$$

$$Y = G + j\omega C.$$

The solution of the differential equations takes the form

$$V = A_1 \varepsilon^{-\sqrt{ZY}\, z} + A_2 \varepsilon^{+\sqrt{ZY}\, z}$$

$$I = \frac{1}{\sqrt{Z/Y}} \left(A_1 \varepsilon^{-\sqrt{ZY}\, z} - A_2 \varepsilon^{+\sqrt{ZY}\, z} \right)$$

or

$$V = A_1 \varepsilon^{-\gamma z} + A_2 \varepsilon^{+\gamma z}$$

$$I = \frac{1}{Z_0} \left(A_1 \varepsilon^{-\gamma z} - A_2 \varepsilon^{\gamma z} \right) \tag{1}$$

where

$$\gamma = \sqrt{ZY} = \sqrt{(R + j\omega L)(G + j\omega C)}$$

is called the propagation constant of the line. The real part of the propagation constant is called the attenuation constant α, and the imaginary part is called the phase constant β. Then

$$\gamma = \alpha + j\beta.$$

The units of the attenuation constant are nepers/meter, and the units of the phase constant are radians/meter.

The characteristic impedance is defined by

$$Z_0 = \sqrt{\frac{Z}{Y}} = \sqrt{\frac{(R+j\omega L)}{(G+j\omega C)}}$$

A lossless, or dissipationless, transmission line is, by definition, a line on which the series resistance and the shunt conductance are negligible, that is, $R = G = 0$. For the lossless case, the propagation constant is imaginary and is given by:

$$\gamma = j\beta = j\omega\sqrt{LC} .$$

The phase constant is

$$\beta = \omega\sqrt{LC} \text{ radians/meter.}$$

The characteristic impedance of a lossless line is real-valued and given by

$$Z_0 = \sqrt{\frac{L}{C}} \text{ ohms.}$$

2. LINE DRIVEN AT ONE END

The schematic representation of a two-conductor transmission line driven at one end and terminated at the other end is shown in Fig. 3. The line is driven by a source with an open circuit voltage V_g and an internal

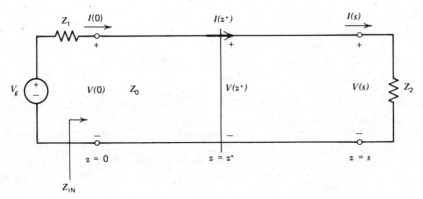

FIGURE 3. Schematic of two-wire line driven at one end.

impedance Z_1, and is terminated in an impedance Z_2. The length of the line is s. The voltage, current, and impedance at the input (or sending end) of the line are denoted by $V(0)$, $I(0)$ and Z_{IN}.

When equations (1) are solved by the usual methods (and the exponential notation is converted to hyperbolic form), the resulting expressions for the voltage, current and impedance at any point z^* along the line are

$$V(z^*) = V(0)\left(\cosh \gamma z^* - \frac{Z_0}{Z_{IN}} \sinh \gamma z^*\right) \tag{2}$$

$$I(z^*) = I(0)\left(\cosh \gamma z^* - \frac{Z_{IN}}{Z_0} \sinh \gamma z^*\right) \tag{3}$$

$$Z(z^*) = \frac{V(z^*)}{I(z^*)} = Z_0 \frac{Z_2 + Z_0 \tanh \gamma(s - z^*)}{Z_0 + Z_2 \tanh \gamma(s - z^*)}. \tag{4}$$

Note that the expressions for the voltage and current are given in terms of the sending-end quantities $V(0)$, $I(0)$ and Z_{IN}. We wish to derive expressions for the voltage and current along the line (i.e., the voltage and current distributions) in terms of the source voltage V_g, source impedance Z_1, and load impedance.

We start with the current distribution given by (3). The sending-end current, referring to Fig. 3, is just

$$I(0) = \frac{V_g}{Z_{IN} + Z_1}.$$

Substituting this into equation (3) yields

$$I(z^*) = V_g \left[\frac{\cosh \gamma z^*}{Z_{IN} + Z_1} - \frac{\sinh \gamma z^*}{Z_0 \left(1 + \dfrac{Z_1}{Z_{IN}}\right)}\right]. \tag{5}$$

The input impedance is obtained directly from (4) with $Z^* = 0$. We have

$$Z_{IN} = Z_0 \frac{Z_2 + Z_0 \tanh \gamma s}{Z_0 + Z_2 \tanh \gamma s}. \tag{6}$$

The denominator of the first term in equation (5) is

$$Z_{IN} + Z_1 = \frac{Z_0 Z_2 + Z_0^2 \dfrac{\sinh \gamma s}{\cosh \gamma s}}{Z_0 + Z_2 \dfrac{\sinh \gamma s}{\cosh \gamma s}} + Z_1$$

or

$$Z_{IN} + Z_1 = \frac{Z_0 Z_2 \cosh \gamma s + Z_0^2 \sinh \gamma s + Z_0 Z_1 \cosh \gamma s + Z_1 Z_2 \sinh \gamma s}{Z_0 \cosh \gamma s + Z_2 \sinh \gamma s}. \tag{7}$$

The denominator of the second term in equation (5), using Z_{IN} from (6), is

$$Z_0 \left(1 + \frac{Z_1}{Z_{IN}}\right) = Z_0 + \frac{Z_1 Z_0 + Z_1 Z_2 \dfrac{\sinh \gamma s}{\cosh \gamma s}}{Z_2 + Z_0 \dfrac{\sinh \gamma s}{\cosh \gamma s}}$$

or

$$Z_0 \left(1 + \frac{Z_1}{Z_{IN}}\right) = \frac{Z_0 Z_2 \cosh \gamma s + Z_0^2 \sinh \gamma s + Z_0 Z_1 \cosh \gamma s + Z_1 Z_2 \sinh \gamma s}{Z_2 \cosh \gamma s + Z_0 \sinh \gamma s}. \tag{8}$$

Substituting (7) and (8) into (5) yields

$$I(z^*) =$$

$$V_g \left[\frac{(Z_0 \cosh \gamma s + Z_2 \sinh \gamma s) \cosh \gamma z^* - (Z_2 \cosh \gamma s + Z_0 \sinh \gamma s) \sinh \gamma z^*}{(Z_0 Z_1 + Z_0 Z_2) \cosh \gamma s + (Z_0^2 + Z_1 Z_2) \sinh \gamma s} \right]$$

which, on use of the hyperbolic identities

$$\sinh A - B = \sinh A \cosh B - \cosh A \sinh B$$

$$\cosh A - B = \cosh A \cosh B - \sinh A \sinh B$$

reduces to

$$I(z^*) = V_g \frac{Z_0 \cosh \gamma (s - z^*) + Z_2 \sinh \gamma (s - z^*)}{(Z_0 Z_1 + Z_0 Z_2) \cosh \gamma s + (Z_0^2 + Z_1 Z_2) \sinh \gamma s} \tag{9}$$

which is the desired current distribution for a dissipative line driven at one end.

The load currents, that is, the currents in the impedances Z_1 and Z_2, are obtained by setting $z^* = 0$ and $z^* = s$ in (9). We have

$$I(0) = V_g \frac{Z_0 \cosh \gamma s + Z_2 \sinh \gamma s}{(Z_0 Z_1 + Z_0 Z_2) \cosh \gamma s + (Z_0^2 + Z_1 Z_2) \sinh \gamma s} \tag{10}$$

and

$$I(s) = \frac{V_g Z_0}{(Z_0 Z_1 + Z_0 Z_2) \cosh \gamma s + (Z_0^2 + Z_1 Z_2) \sinh \gamma s}. \tag{11}$$

For a lossless line, the current distribution is obtained from (9) with $\gamma = j\beta$. The result is

$$I(z^*) = V_g \frac{Z_0 \cos \beta (s - z^*) + j Z_2 \sin \beta (s - z^*)}{(Z_0 Z_1 + Z_0 Z_2) \cos \beta s + j(Z_0^2 + Z_1 Z_2) \sin \beta s}. \tag{12}$$

Similarly, the load currents on a lossless line driven at one end are

$$I(0) = V_g \frac{Z_0 \cos \beta s + j Z_2 \sin \beta s}{(Z_0 Z_1 + Z_0 Z_2) \cos \beta s + j(Z_0^2 + Z_1 Z_2) \sin \beta s} \tag{13}$$

and

$$I(s) = \frac{V_g Z_0}{(Z_0 Z_1 + Z_0 Z_2) \cos \beta s + j(Z_0^2 + Z_1 Z_2) \sin \beta s}. \tag{14}$$

The voltage distribution may be derived in an analogous manner starting with (2). However, the current distribution has already been derived, and by definition

$$V(z^*) = I(z^*) Z(z^*). \tag{15}$$

The desired voltage distribution for a dissipative line is obtained by substituting (9) and (4) into (15). The result is

$$V(z^*) = V_g \frac{Z_0 Z_2 \cosh \gamma (s - z^*) + Z_0^2 \sinh \gamma (s - z^*)}{(Z_0 Z_1 + Z_0 Z_2) \cosh \gamma s + (Z_0^2 + Z_1 Z_2) \sinh \gamma s}. \tag{16}$$

At the receiving end of the line, the voltage across terminating impedance Z_2 is obtained from (16) with $z^* = s$ or, equivalently, by the current $I(s)$ in (11) times the impedance. That is,

$$V_{Z_2} = V(s) = I(s)Z_2 = \frac{V_g Z_0 Z_2}{(Z_0 Z_1 + Z_0 Z_2)\cosh \gamma s + (Z_0^2 + Z_1 Z_2)\sinh \gamma s}. \quad (17)$$

At the driven end of the line, the voltage across impedance Z_1 is obtained by multiplying the current (10) by the impedance, or (referring to Fig. 3) from the relation $V_g - V(0)$, where $V(0)$ is given by (16) with $Z^* = 0$. The resulting expression is

$$V_{Z_1} = I(s)Z_1 = V_g - V(0)$$

$$= V_g \frac{Z_0 Z_1 \cosh \gamma s + Z_1 Z_2 \sinh \gamma s}{(Z_0 Z_1 + Z_0 Z_2)\cosh \gamma s + (Z_0^2 + Z_1 Z_2)\sinh \gamma s}. \quad (18)$$

For lossless lines, the voltage distribution and load voltages are found by making the substitution $\gamma = j\beta$ in (16), (17), and (18) and by using the identities $\cosh jA = \cos A$ and $\sinh jA = j\sin A$.

3. LINE DRIVEN AT BOTH ENDS

A two-conductor transmission line driven at both ends is shown in Fig. 4. The current and voltage distributions along the line, expressed in terms of the source generators V_{g1} and V_{g2} and source impedances Z_1 and Z_2, are desired.

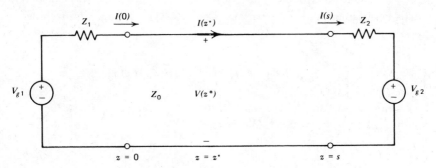

FIGURE 4. Two-wire line driven at both ends.

Again, we begin with the current distribution. The current at any point z^* on the line due to the left-hand source V_{g1} is given by equation (9) as

$$I_1(z^*) = V_{g1} \frac{Z_0 \cosh \gamma(s - z^*) + Z_2 \sinh \gamma(s - z^*)}{(Z_0 Z_1 + Z_0 Z_2) \cosh \gamma s + (Z_0^2 + Z_1 Z_2) \sinh \gamma s}. \tag{19}$$

The current distribution due to the right-hand source V_{g2} follows directly from (19), replacing V_{g1} with V_{g2}, Z_2 with Z_1, and $s - z^*$ with z^*. Then

$$I_2(z^*) = -V_{g2} \frac{Z_0 \cosh \gamma z^* + Z_1 \sinh \gamma z^*}{(Z_0 Z_1 + Z_0 Z_2) \cosh \gamma s + (Z_0^2 + Z_1 Z_2) \sinh \gamma s}. \tag{20}$$

The minus sign results since the current at z^* due to V_{g2} flows in the opposite direction from the current due to V_{g1} for the assigned generator polarities in Fig. 4.

The desired lossy-line current distribution is obtained, by superposition, from the sum of (19) and (20). The result is

$$I(z^*) = \frac{V_{g1}}{D} \left[Z_0 \cosh \gamma(s - z^*) + Z_2 \sinh \gamma(s - z^*) \right]$$

$$- \frac{V_{g2}}{D} \left[Z_0 \cosh \gamma z^* + Z_1 \sinh \gamma z^* \right] \tag{21}$$

where

$$D = (Z_0 Z_1 + Z_0 Z_2) \cosh \gamma s + (Z_0^2 + Z_1 Z_2) \sinh \gamma s.$$

(One will note that the denominator function D recurs frequently enough in transmission line equations to deserve its own symbol).

The voltage distribution for a lossy two-conductor line driven at both ends is obtained by superposition using (16). The voltage at any point z^* due to V_{g1} is

$$V_1(z^*) = \frac{V_{g1}}{D} \left[Z_0 Z_2 \cosh \gamma(s - z^*) + Z_0^2 \sinh \gamma(s - z^*) \right]. \tag{22}$$

The voltage due to V_{g2} is found from (22) by replacing V_{g1} with V_{g2}, Z_2 with Z_1, and $s - z^*$ with z^*. We have

$$V_2(z^*) = \frac{V_{g2}}{D} \left[Z_0 Z_1 \cosh \gamma z^* + Z_0^2 \sinh \gamma z^* \right]. \tag{23}$$

Summing equations (22) and (23) yields the desired voltage distribution. (Note that although the indicated generator polarities in Fig. 4 produce currents which flow in opposite directions, the voltages have the same sense on the line). We have

$$V(z^*) = \frac{V_{g1}}{D} \left[Z_0 Z_2 \cosh \gamma (s - z^*) + Z_0^2 \sinh \gamma (s - z^*) \right]$$

$$+ \frac{V_{g2}}{D} \left[Z_0 Z_1 \cosh \gamma z^* + Z_0^2 \sinh \gamma z^* \right]. \tag{24}$$

In general, the generators V_{g1} and V_{g2} in Fig. 4 and in equations (21) and (24) are completely independent. However, two cases are of special interest. The first case is when both generators are identical both in amplitude and phase, that is, $V_{g1} = V_{g2} = V_g$. The current distribution from (21) becomes

$$I(z^*) = \frac{V_g}{D}$$

$$\times \left[Z_0 \cosh \gamma (s - z^*) - Z_0 \cosh \gamma z^* + Z_2 \sinh \gamma (s - z^*) - Z_1 \sinh \gamma z^* \right]. \tag{25}$$

The load currents are

$$I(0) = \frac{V_g}{D} \left[Z_2 \sinh \gamma s - Z_0 (1 - \cosh \gamma s) \right] \tag{26}$$

$$I(s) = \frac{V_g}{D} \left[Z_0 (1 - \cosh \gamma s) - Z_1 \sinh \gamma s \right]. \tag{27}$$

For a lossless line with $V_{g1} = V_{g2} = V_g$, the current distribution and load currents are given by

$$I(z^*) = \frac{V_g}{D} \left[Z_0 \cos \beta (s - z^*) - Z_0 \cos \beta z^* + j Z_2 \sin \beta (s - z^*) - j Z_1 \sin \beta z \right] \tag{28}$$

$$I(0) = \frac{V_g}{D} \left[j Z_2 \sin \beta s - Z_0 (1 - \cos \beta s) \right] \tag{29}$$

$$I(s) = \frac{V_g}{D} \left[Z_0 (1 - \cos \beta s) - j Z_1 \sin \beta s \right] \tag{30}$$

where now for the lossless case

$$D = (Z_0 Z_1 + Z_0 Z_2) \cos \beta s + j(Z_0^2 + Z_1 Z_2) \sin \beta s. \tag{31}$$

The voltage distribution on a lossy line for the case when both driving generators are identical is

$$V(z^*) = \frac{V_g Z_0}{D} \Big[Z_2 \cosh \gamma (s - z^*) + Z_1 \cosh \gamma z^*$$

$$+ Z_0 \sinh \gamma (s - z^*) + Z_0 \sinh \gamma z^* \Big] \tag{32}$$

where

$$D = (Z_0 Z_1 + Z_0 Z_2) \cosh \gamma s + (Z_0^2 + Z_1 Z_2) \sinh \gamma s. \tag{33}$$

On a lossless line, the voltage distribution is

$$V(z^*) = \frac{V_g Z_0}{D} \Big[Z_2 \cos \beta (s - z^*) + Z_1 \cos \beta z^*$$

$$+ j Z_0 \sin \beta (s - z^*) + j Z_0 \sin \beta z^* \Big] \tag{34}$$

where D is given by (31).

From the current distribution equation (25), the current at the midpoint of the line ($z^* = s/2$) for the case when both generators are identical is

$$I(s/2) = \frac{V_g}{D}(Z_2 - Z_1) \sinh \frac{\gamma s}{2}.$$

When $Z_1 = Z_2$, the current at the midpoint is identically zero. This is the expected result. The voltage at the midpoint under these conditions is maximum and is equal to

$$V(s/2) = \frac{V_g}{Z_0} \varepsilon^{-j\beta s/2}.$$

The second case of interest is when V_{g1} and V_{g2} have identical amplitudes but are displaced in phase by βs radians (the delay encountered by a wave traveling the length of the line). That is,

$$V_{g1} = V_g$$
$$V_{g2} = V_g \varepsilon^{-j\beta s}. \tag{35}$$

Substituting (35) into the lossless line current distribution derived from equation (21) yields

$$I(z^*) = \frac{V_g}{D}\left[Z_0\cos\beta(s-z^*)+jZ_2\sin\beta(s-z^*)\right]$$

$$-\frac{V_g}{D}\left[Z_0\cos\beta z^*+jZ_1\sin\beta z^*\right]\varepsilon^{-j\beta s}$$

where D is given by (31).

Using the identity

$$\varepsilon^{-j\beta s} = \cos\beta s - j\sin\beta s$$

and carrying out the mathematical manipulations gives the following expression for the current distribution:

$$I(z^*) = \frac{V_g}{D}\left[(Z_0-Z_1)\sin\beta s\sin\beta z^* + j(Z_0+Z_2)\sin\beta s\cos\beta z^*\right.$$

$$\left.-j(Z_1+Z_2)\cos\beta s\sin\beta z^*\right]. \tag{36}$$

The load currents are

$$I(0) = \frac{V_g}{D}(Z_0+Z_2)\sin\beta s \tag{37}$$

$$I(s) = \frac{V_g}{2D}(Z_0-Z_1)\left[(1-\cos 2\beta s)+j\sin 2\beta s\right]. \tag{38}$$

4.   LINE DRIVEN IN THE MIDSECTION BY A SINGLE SOURCE

A two-conductor transmission line with a single voltage source $V_g(z)$ located at a point z in the midsection of the line is shown in Fig. 5. It is desired to find the current at any point z^* on the line, that is, the current distribution $I(z^*)$.

The current distribution to the right of the source $(z^* > z)$ is derived by replacing the section of line to the left of the source by the impedance Z_{IN}^L looking into that section as shown in Fig. 6. The impedance looking to the left, referring to equation (6), can be written

$$Z_{IN}^L = Z_0\frac{Z_1+Z_0\tanh\gamma z}{Z_0+Z_1\tanh\gamma z} \tag{39}$$

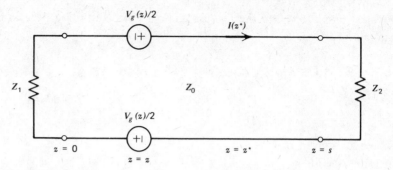

FIGURE 5. Two-conductor line with a source in the midsection.

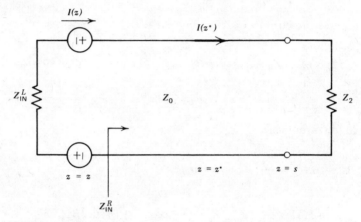

FIGURE 6. Equivalent line for deriving current distribution when $z^* > z$.

since the distance to the left-hand termination Z_1 is z.

Similarly, the impedance looking to the right of the source is

$$Z_{\text{IN}}^R = Z_0 \frac{Z_2 + Z_0 \tanh \gamma (s - z)}{Z_0 + Z_2 \tanh \gamma (s - z)}. \tag{40}$$

On inspection of Fig. 6, the current on the line at point z can be written

$$I(z) = \frac{V_g(z)}{Z_{\text{IN}}^L + Z_{\text{IN}}^R} \tag{41}$$

Equation (3) gives the current distribution on a line in terms of the sending end current $I(0)$. Making the appropriate shift in coordinates yields

$$I(z^*) = I(z)\left[\cosh\gamma(z^* - z) - \frac{Z_{IN}^R}{Z_0}\sinh\gamma(z^* - z)\right]. \tag{42}$$

The desired expression for the current distribution is found by substituting (39), (40), and (41) in (42). The result, after a great deal of mathematical manipulation, is

$$I(z^*) = \frac{V_g(z)}{Z_0 D}\left[Z_0\cosh\gamma z + Z_1\sinh\gamma z\right]$$

$$\times\left[Z_0\cosh\gamma(s - z^*) + Z_2\sinh\gamma(s - z^*)\right] \qquad z^* > z$$

and (43)

$$I(z^*) = \frac{V_g(z)}{Z_0 D}\left[Z_0\cosh\gamma z^* + Z_1\sinh\gamma z^*\right]$$

$$\times\left[Z_0\cosh\gamma(s - z) + Z_2\sinh\gamma(s - z)\right] \qquad z^* < z$$

where

$$D = (Z_0 Z_1 + Z_0 Z_2)\cosh\gamma s + (Z_0^2 + Z_1 Z_2)\sinh\gamma s.$$

The expression for the current distribution to the left of the source $(z^* < z)$ given in (43) is obtained in a manner analogous to that indicated for $z^* > z$.

Evaluating (43) at $z^* = s$ and $z^* = 0$ yields the load currents. We have

$$I(s) = \frac{V_g(z)}{D}\left[Z_0\cosh\gamma z + Z_1\sinh\gamma z\right] \tag{44}$$

$$I(0) = \frac{V_g(z)}{D}\left[Z_0\cosh\gamma(s - z) + Z_2\sinh\gamma(s - z)\right]. \tag{45}$$

When $z = 0$, that is, when the source is at the left-hand end of the line, equations (43), (44), and (45) reduce to the corresponding expressions for a line driven at the end given by (9), (10), and (11).

The current distribution and the load currents for a lossless line driven by a single source in the midsection are

$$I(z^*) = \frac{V_g(z)}{Z_0 D} \left[Z_0 \cos \beta z + j Z_1 \sin \beta z \right]$$

$$\times \left[Z_0 \cos \beta (s - z^*) + j Z_2 \sin \beta (s - z^*) \right] \quad z^* > z$$

and

$$I(z^*) = \frac{V_g(z)}{Z_0 D} \left[Z_0 \cos \beta z^* + j Z_1 \sin \beta z^* \right]$$

$$\times \left[Z_0 \cos \beta (s - z) + j Z_2 \sin \beta (s - z) \right] \quad z^* < z$$

and

$$I(s) = \frac{V_g(z)}{D} \left[Z_0 \cos \beta z + j Z_1 \sin \beta z \right]$$

$$I(0) = \frac{V_g(z)}{D} \left[Z_0 \cos \beta (s - z) + j Z_2 \sin \beta (s - z) \right]$$

where, for the lossless case,

$$D = (Z_0 Z_1 + Z_0 Z_2) \cos \beta s + j (Z_0^2 + Z_1 Z_2) \sin \beta s.$$

5.   LINE WITH DISTRIBUTED SOURCES ALONG THE CONDUCTORS

Let $K(z)$ be a continuous distribution of voltage sources along the line as illustrated in Fig. 7. In general, $K(z)$ can vary as a function of position along the line. The dimensions of $K(z)$ are volts per meter (the same as electric field strength).

The incremental voltage ΔV over a small section of line of length Δz where K is approximately constant is $\Delta V = K \Delta z$. In the limit as Δz approaches zero, we have

$$dV(z) = K(z) dz. \tag{46}$$

The current at any point z^* on the line is found by substituting (46) for $V_g(z)$ in (43) and summing the contributions from all the $dV(z)$ sources,

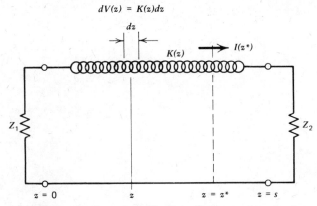

FIGURE 7. Continuous distribution of sources along the line.

that is, integrating over the length of the line. The result is

$$I(z^*) = \frac{Z_0 \cosh \gamma (s - z^*) + Z_2 \sinh \gamma (s - z^*)}{Z_0 D}$$

$$\times \int_0^{z^*} K(z) \left[Z_0 \cosh \gamma z + Z_1 \sinh \gamma z \right] dz$$

$$+ \frac{Z_0 \cosh \gamma z^* + Z_1 \sinh \gamma z^*}{Z_0 D}$$

$$\times \int_{z^*}^s K(z) \left[Z_0 \cosh \gamma (s - z) + Z_2 \sinh \gamma (s - z) \right] dz$$

where

$$D = (Z_0 Z_1 + Z_0 Z_2) \cosh \gamma s + (Z_0^2 + Z_1 Z_2) \sinh \gamma s.$$

The load currents are found by substituting $z^* = 0$ and $z^* = s$ in the foregoing current distribution. We have

$$I(0) = \frac{1}{D} \int_0^s K(z) \left[Z_0 \cosh \gamma (s - z) + Z_2 \sinh \gamma (s - z) \right] dz$$

and

$$I(s) = \frac{1}{D} \int_0^s K(z) \left[Z_0 \cosh \gamma z + Z_1 \sinh \gamma z \right] dz.$$

BIBLIOGRAPHY

CHAPTER 1   NONUNIFORM FIELDS

[1] C. D. Taylor, R. S. Satterwhite, and C. W. Harrison, "The Response of a Termir Two-Wire Transmission Line Excited by a Nonuniform Electromagnetic Field," *I Trans. on Antennas and Propagation*, Vol. AP-13, No. 6, pp. 987–989, November, 196

[2] A. A. Smith, Jr., "A More Convenient Form of the Equations for the Response Transmission Line Excited by Nonuniform Fields," *IEEE Trans. on EMC*, Vol. EMC No. 3, pp. 151–152, August 1973.

[3] C. Whitescarver, "Transient Electromagnetic Field Coupling with Two-Wire Unif Transmission Lines," Ph.D dissertation, University of Florida, Gainesville, March, 1

CHAPTER 2   PLANE WAVES

[1] R. W. P. King and C. W. Harrison, Jr., "Transmission Line Coupled to a Cylinder ir Incident Field," *IEEE Trans. on EMC*, Vol. EMC-14, No. 3 pp. 97–105, August, 197%

[2] C. W. Harrison, Jr., "The Response of a Terminated Two-Wire Line Suspended in above a Semi-Infinite Dissipative Medium and Excited by a Plane-Wave RF F: Generated in Free Space," *IEEE Trans. on EMC*, Vol. EMC-11, No. 4, pp. 149–1 November, 1969.

[3] C. W. Harrison, Jr. and M. L. Houston, "The Response of a Terminated Two-Wire L Buried in the Earth and Excited by a Plane-Wave RF Field Generated in Free Spac *IEEE Trans. on EMC*, Vol. EMC-11, No. 4, pp. 144–148, November, 1969.

[4] C. W. Harrison, Jr., "Generalized Theory of Impedance Loaded Multiconductor Tra mission Lines in an Incident Field," *IEEE Trans. on EMC*, Vol. EMC-14, No. 2, j 56–63, May, 1972.

[5] C. W. Harrison, Jr., "Bounds on the Load Currents of Exposed One- and Two-Conduc Transmission Lines Electromagnetically Coupled to a Rocket," *IEEE Trans. on EM* Vol. EMC-14, No. 1, pp. 4–9, February, 1972.

[6] R. W. P. King and C. W. Harrison, Jr., "Excitation of an External Terminated Longitut nal Conductor on a Rocket by a Transverse Electromagnetic Field," *IEEE Trans. EMC*, Vol. EMC-14, No. 1, pp. 1–3, February, 1972.

[7] C. W. Harrison, Jr., "Receiving Characteristics of Two-Wire Lines Excited by Unifor and Non-uniform Electric Fields," Sandia Corp., Report No. SC-R-64-164, May, 1964.

CHAPTER 3   NONUNIFORM FIELDS OF LOOP AND DIPOLE

[1] S. A. Schelkunoff and H. T. Friis, *Antennas Theory and Practice*, New York, Londor Sidney: John Wiley & Sons, 1952, pp. 319–320.

[2] J. D. Kraus, *Antennas*, New York, Toronto, London: McGraw-Hill Book Co., Inc., 1950, pp. 127–137.

[3] A. A. Smith, Jr., "The Response of a Two-Wire Transmission Line Excited by the Nonuniform Electromagnetic Fields of a Nearby Loop," *IEEE Trans. on EMC*, Vol. EMC-16, No. 4, pp. 196–200, November, 1974.

[4] A. A. Smith, Jr., "Load Current Spectrum of a Two-Wire Transmission Line Excited by a Nearby Dipole," Proceedings of the Electromagnetic Compatibility Symposium, Montreux, May 20–22, 1975.

CHAPTER 4   SHIELDED CABLES

[1] S. Shenfeld, "Coupling Impedance of Cylindrical Tubes," *IEEE Trans. on EMC*, Vol. EMC-14, No. 1, pp. 10–16, February, 1972.

[2] D. A. Miller and P. P. Toulios, "Penetration of Coaxial Cables by Transient Fields," *1968 IEEE EMC Symposium Record*, pp. 414–423.

[3] R. M. Whitman, "Cable Shielding Performance and CW Response," *IEEE Trans. on EMC*, Vol. EMC-15, No. 4, pp. 180–187, November, 1973.

[4] C. N. McDowell and M. J. Bernstein, "Surface Transfer Impedance Measurements on Subminiature Coaxial Cables," *IEEE Trans. on EMC*, Vol. EMC-15, No. 4, pp. 188–190, November, 1973.

[5] A. A. Smith, Jr., "Electric Field Propagation in the Proximal Region," *IEEE Trans. on EMC*, Vol. EMC-11, No. 4, pp. 151–163, November, 1969.

INDEX

DATE DUE

~~MAY 12~~			
~~JUN 14~~			
~~JUL 07~~			
~~			
~~AUG~~			
MAR 7 '93			
MAR 17 '93			
GAYLORD			PRINTED IN U.S.A

COUPLING OF EXTERNAL ELECTROMAGNETIC FIE
TK 3221 S58